OpenSCAD Cookbook

OpenSCAD Recipes for learning 3D modeling

by

John Clark Craig

A Books To Believe In Publication
All Rights Reserved

Copyright 2018 by John Clark Craig

No part of this book may be reproduced or transmitted in any form or by any means, electronic or mechanical, including photocopy, recording or by any information storage and retrieval system, without permission, in writing from the publisher.

Proudly Published in the USA by
Books To Believe In

Phone: (303) 794-8888

OpenSCADbook.com
BooksToBelieveIn.com

First Edition: ISBN: 9781790273911

Dedication

To my mother,

who sparked my curiosity
and opened my mind.

For our many talks through the years on the many mysteries of the Universe.

For always being a source of inspiration.

Table of Contents

Getting Started   9
   Why Use OpenSCAD
   Install OpenSCAD
   Cheatsheet
   How to Learn from this Book

Recipe 1: **Hello World Meatball!**   15
Recipe 2: **Create a Square Sheetcake**   21
Recipe 3: **Parameterization**   29
Recipe 4: **Create a Circle**   35
Recipe 5: **Rotation and Translation**   41
Recipe 6: **Create a Polygon**   51

Recipe 7: **Trimming the Edges**   57
Recipe 8: **Stamp Your Name On It**   63
Recipe 9: **Extruding Into Space**   69
Recipe 10: **Create a Donut**   81
Recipe 11: **Kitchen Tips and Tricks**   89
Recipe 12: **Functions, Modules, and Regular Polygons**   97

Recipe 13: **No Matter How You Slice It**	105
Recipe 14: **Create the "Holey" Grail**	113
Recipe 15: **Birthday Candles & Other Common Cylinders**	123
Recipe 16: **Ice Cubes for Party Drinks**	131
Recipe 17: **Polyhedron Souffle**	139
Recipe 18: **After-Dinner Mints and Toothpicks**	145
Recipe 19: **Use a Recipe Box**	151
Recipe 20: **Mirror Mirror on the Plane**	159
Recipe 21: **Popcorn and Other Hulls**	167
Recipe 22: **Minkowski Mints**	179

Appendix A 187
Using OpenSCAD
    Menus
    Icons and Buttons
    Mouse Use
    Creating STL Files
    Animation

Index 196

About John Clark Craig 201

OpenSCAD Cookbook

Getting Started

The best way to learn OpenSCAD is to jump right in and start using it. This introduction explains many of the basic syntax rules and other details of creating 3D objects by creating simple shapes and objects, with side explanations to help you firmly grasp the concepts.

You can't really learn to ride a bicycle by reading a book or watching an online video. You need to actually go outside, get in the saddle, and try riding. With a little effort, you'll quickly learn to get balanced and be a true biker. It's the same with OpenSCAD. As you read these recipes, go ahead and fire up OpenSCAD to try them. Change the numbers, experiment freely, see if you can predict what will happen if you make little changes here and there. You'll become a true 3D modeling guru much faster this way.

Above all, have some fun. There's something magical about typing a line or two of commands and suddenly seeing a 3D object appear in space, ready to rotate, zoom, and pan to get a solid feel for its, well, solidness. It's very cool. I've couched the content in this book into "recipes" to help keep things interesting. Just like with a cookbook, try adjusting the recipes a little here and there to see what happens. Also, I might throw in a food pun or two. I can't help it.

Why use OpenSCAD

The main difference between OpenSCAD and other interactive 3D modeling software packages is in the way you create the 3D objects. Most software uses the mouse and keyboard to draw on the fly, with little to no text input. OpenSCAD on the other hand lets you type out explicit text commands, as a script or program, that drive the creation of the objects. In both types of programs you get to see the results, including spinning, scaling, panning, and zooming the view, but the way you create the objects is quite different.

I suggest trying both types of 3D modeling software to see what works best for the way your own brain works. For me, OpenSCAD was by far the better way to work, as it lets me have complete understanding of the effects of changes and modifications as I go. Of course your mileage may vary.

Be aware that OpenSCAD is great for mechanical design, but not so great for what I call artistic 3D design. If your goal is to create usable objects on a 3D printer, even complex systems of parts, OpenSCAD is a great choice. If you are looking to create things like 3D animated figures for gaming, or complex organic shapes of a generally non-mathematical nature, there are other software tools that will work better for you, such as Blender.

Install OpenSCAD

If you haven't installed OpenSCAD yet, go to http://www.openscad.org and look for the download instructions for your operating system. You can use OpenSCAD in Windows, on your Mac, or in Linux. The Windows installation is very easy, very quick, and with very low impact on your system. Of course a decently fast graphics card will help things run better and faster.

Cheat Sheet

Documentation for OpenSCAD is a little hodgepodge, and the quality varies a lot. My favorite starting point for researching features of the language is the official "cheat sheet" page that has links into the primary documentation for each command, key word, and system variable.

http://www.openscad.org/cheatsheet

While trying out the following recipes have this link handy so you can get detailed information when desired.

How to Learn from this Book

Learning by doing is what this book is all about. Each recipe adds a few more knowledge bits to your understanding of OpenSCAD, presented in small bites for easy digestion. Each recipe is as short as possible, without being too short, to make it easier for you to type in, or copy and paste if desired, so you can actually interact with the code and try out new things. Skim through the recipes sequentially, and when you see something interesting, or something even the least bit mysterious to you, then by all means play with that recipe for a while.

I started my programming career with the BASIC programming language, eventually writing a bunch of popular books for Microsoft Press and O'Reilly Media. BASIC let you try things out real quick, with the result that learning and absorbing all the syntax and language details happened fast and easy. OpenSCAD has that same great nature to it. Type in a few lines from these recipes, try to guess what will happen when you change some detail, and THEN give it a try. I guarantee you'll learn very fast, while having a lot of fun, and in no time at all you'll be 3D printing some fun, productive, and amazing stuff. Be the guru!

I hope you like the flavor of these recipes!

Notes

OpenSCAD Cookbook

Recipe 1

Hello World Meatball!

Just about every modern programming language has "Hello World!" as its very simplest example program. This is useful to make sure you have the language up and running correctly, that you are entering text or commands correctly, and to basically get the basics out of the way. You know… did you preheat the oven?

For this recipe type in the following one-line command, making sure to add the semicolon at the end of the command:

```
sphere(10);
```

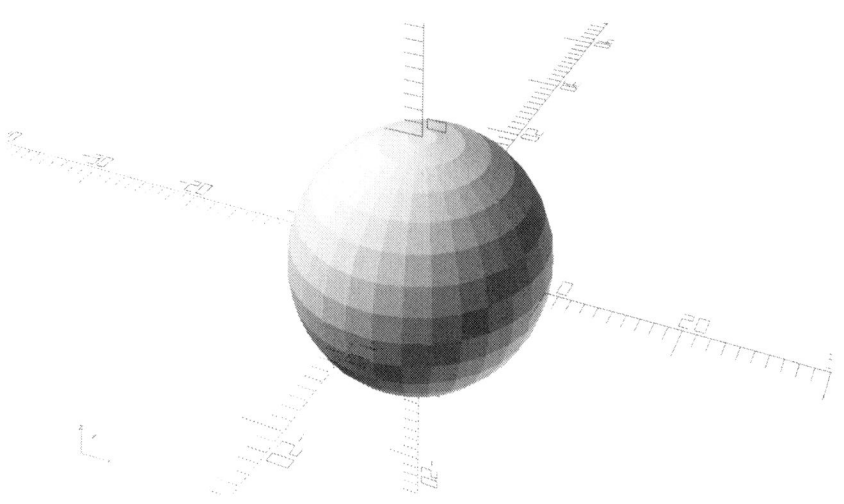

This doesn't display the text "Hello World!" like most computer languages do, but this spherical globe looks like a world in space, or perhaps a tasty meatball, so in a way it's much more cool.

If this is the first time you've created something in OpenSCAD, now is a great time to get familiar with the menus, editing window, and the view window. Rather than get sidetracked from our recipes, Appendix A covers the user interface in detail. If you need a refresher on how to spin, pan, or zoom the sphere, check out Appendix A.

Now let's explore some finer details around the sphere() command in the OpenSCAD language.

You can name the default radius parameter with an "r=" if desired, or with "d=" if you want to provide the diameter instead of the radius. The following three commands all create the same sphere as shown above:

```
sphere(10);
sphere(r=10);
sphere(d=20);
```

Many of OpenSCAD's commands have named parameters, but usually the default parameters without names are simpler and end up being used most often. Just be aware that some commands have extra parameters that might provide some functionality you'll need some day.

You probably noticed the sphere is comprised of many small, flat rectangles, and is not very smooth at all. The default resolution for creating curved surfaces is a compromise between speed and smoothness. It works well using the default setting, but it's easy to change this setting for smoother surfaces. A special variable named $fn can be used to set the "number of fragments" for creating curved surfaces. There are other special variables at play here as well, but in almost all cases just setting $fn will accomplish what you need.

```
$fn = 64;
sphere(10);
```

Setting $fn to 30 creates the same resolution sphere rendering as not setting the $fn value at all. Any number greater than 30 causes more rectangles to be drawn, creating a smoother appearance. Setting $fn to numbers less than 30 creates a rougher sphere surface. As an interesting experiment, try setting $fn to 4, 5, or other small numbers.

Try these commands to see an even smoother meatball... I mean sphere:

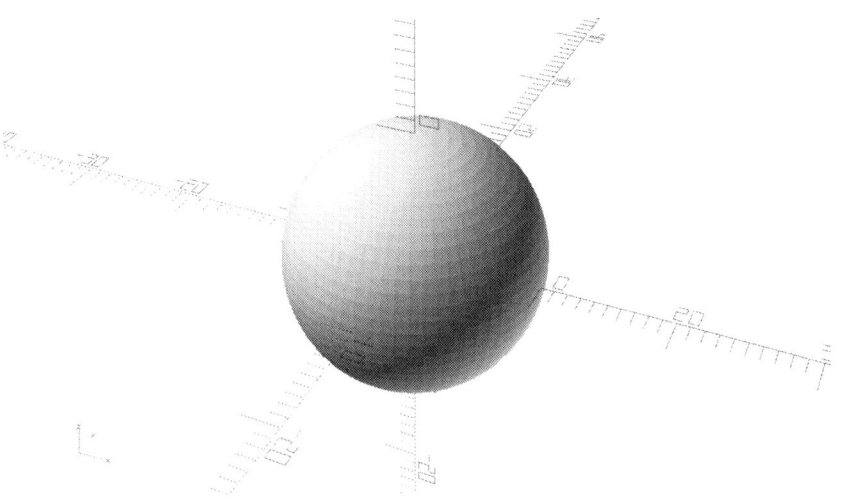

Note that the OpenSCAD manual suggests not setting $fn greater than 100, because it causes your computer to work extra long and hard to render the spherical images. I've used $fn set to 512 without any hassles on my system, but experiment cautiously with larger numbers, and be patient if nothing seems to be happening. Behind the scenes your computer is likely just working overtime to get all the geometry created.

By the way, at 512 or somewhat less, the sphere appears perfectly smooth! Here's how it renders on my Windows 10 computer:

```
$fn = 512;
sphere(10);
```

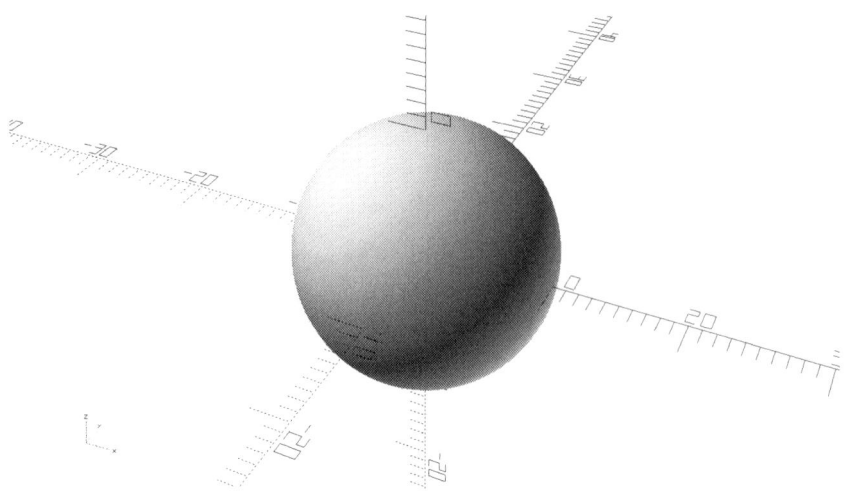

The OpenSCAD manual also suggests setting $fn to numbers that are an exact multiple of 4, for technical reasons that you can read about there, if you're interested. I usually use numbers like 32 or 64 or even 512 just out of habit for this reason, but other numbers work okay in almost all cases.

While developing objects it's a good idea to leave $fn at or near its default value of 30 to render quickly while retaining plenty of resolution to be able to see the "smooth" surfaces accurately. Also note that the goal is often to create an STL file (see Appendix) for handing off to a 3D printer, and $fn will affect the size of the file and the time it will take to print.

Notes

OpenSCAD Cookbook

Recipe 2

Create a Square Sheetcake

Enter the following command to create a simple square in the x,y plane:

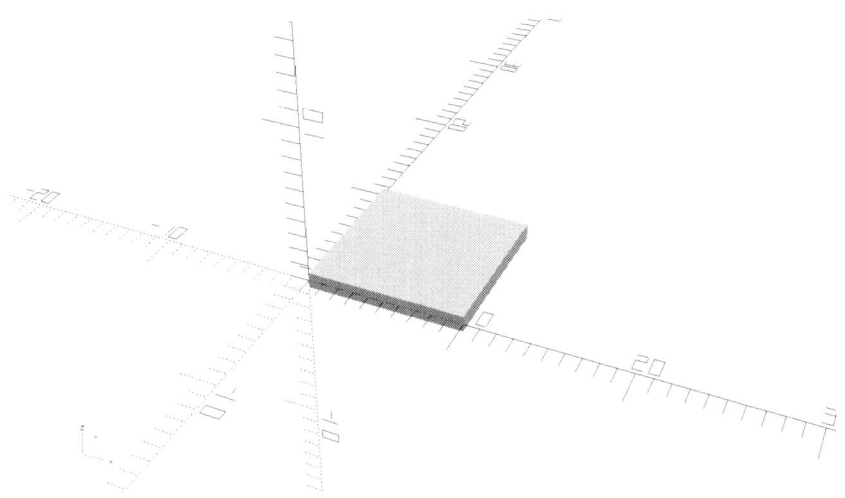

Okay, since OpenSCAD is a 3D modeling program, why are we drawing a "square" instead of a rectangular solid? Actually, the handful of 2D objects that OpenSCAD creates (circle, square, polygon, and text) are each rendered as thin 3D slabs with a thickness of 1 unit... not much flavor there. Mathematically, and internal to the program, they are flat and 2D, but for ease of viewing they are rendered with a tiny bit of thickness, always at 1 unit, like a very thin mint.

Of course "tiny bit" depends on just how big you create your square. A square of size 1000 looks very thin, whereas a square of size 1 looks like a cube. You'll have to zoom in or out to see these effectively, but do give them a try:

```
square(1000);
```

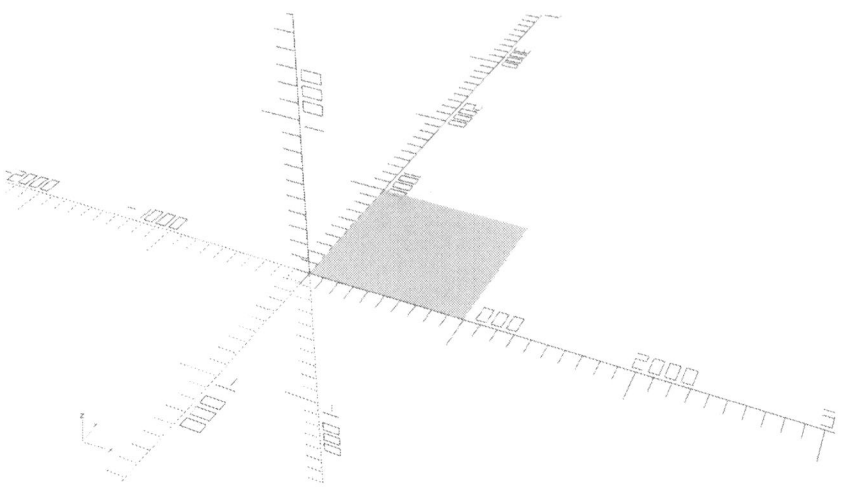

```
square(1);
```

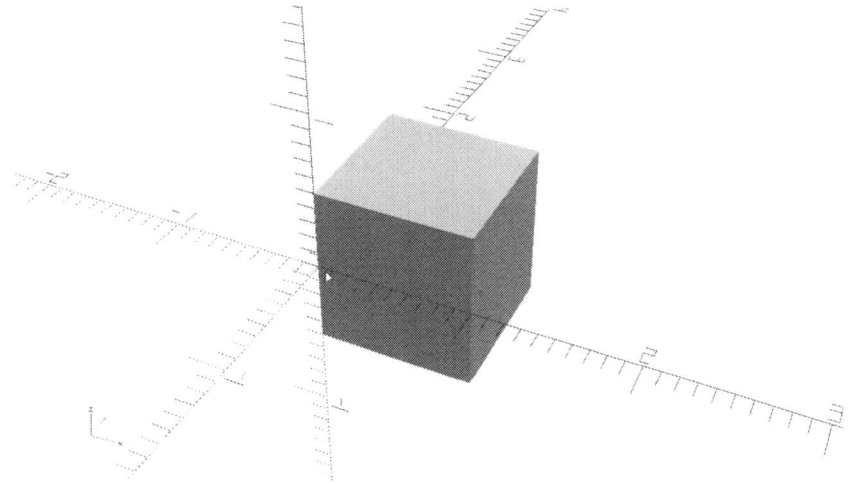

Notice that the 1 unit of thickness is split half above the X,Y plane and half below. The top surface is at 0.5 units on the y-axis, and the bottom surface is at -0.5 units. Keep in mind that the square is considered flat and is simply rendered with this thickness for easier viewing. Later, we'll "extrude" these flat objects to give them real thickness for 3D printing.

A commonly used parameter of the square() command is "center = true". This translates the square, no matter what size, such that its center is at the origin point (0,0,0):

```
square(10, center=true);
```

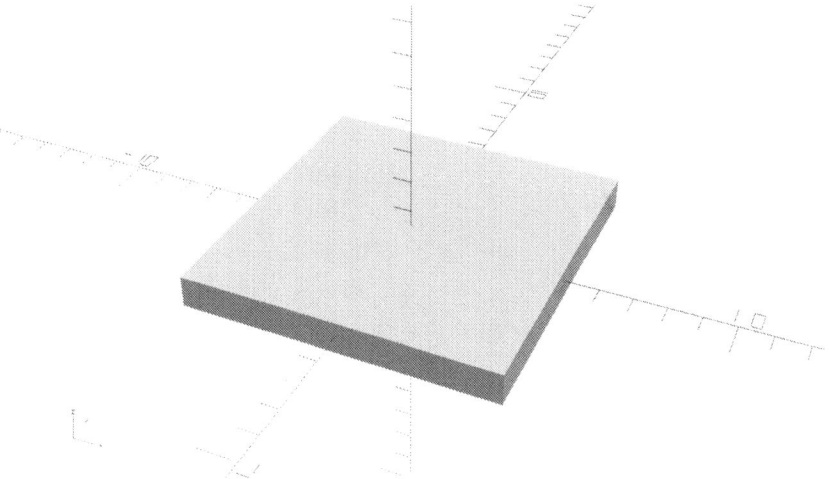

Note the edges are now at 5 units from the origin. The square has not changed in size, only in location.

Create a rectangle

If there was a "rectangle" command in OpenSCAD (there isn't), I would present a separate recipe showing how to use it. However, a slight modification of the square() command provides the way to create 2D rectangles. Instead of a size number, give your square command a list of two numbers to set the rectangle's width and length.

Similar to a list in the Python programming language, a list in OpenSCAD is simply a sequence of numbers enclosed in square brackets, in this case a list of two numbers. These lists in OpenSCAD are often described as "vectors", but I prefer to call them lists, mostly because these lists carry a variety of types of information, depending on context.

For example, to create a rectangle 10 units wide (along the x-axis), and 5 units long (along the y-axis) use the list [10,5]. Type in this modification of the square command to see how this works:

```
square([10,5]);
```

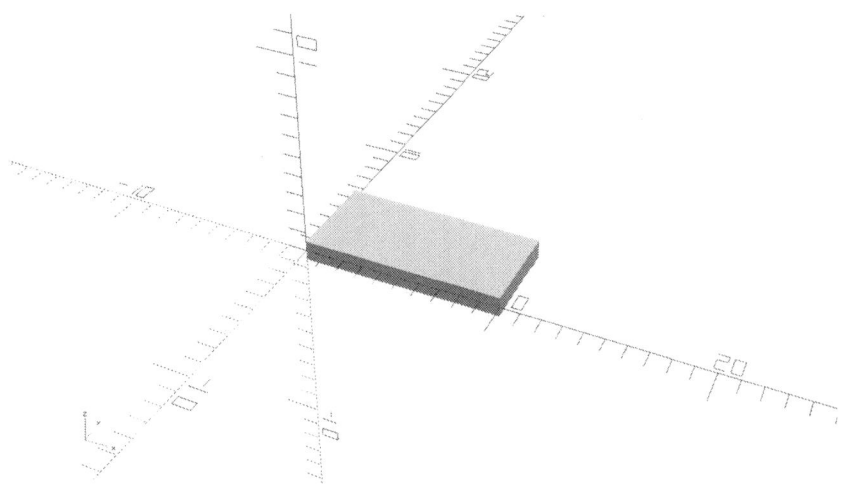

Another way to create the exact same rectangle is to create a size variable first, and then pass that variable to the square command. This is one way to make your programs more readable and more flexible, by isolating calculations outside the action commands. Here are three sets of commands that all create exactly the same rectangle as above:

```
rect_size=[10,5];
square(rect_size);


wide=10;
size=[wide,5];
square(size);


wide=10; high=5;
size=[wide,high];
square(size);
```

In the last example, two variables are assigned values on the same line of text. This is an okay practice in OpenSCAD, and some people like to group commands in this way. For clarity, I tend to separate commands on their own lines, and you'll see this technique used throughout this book, but feel free to structure your own programs in a way that works for you.

Notes

Notes

OpenSCAD Cookbook

Recipe 3

Parameterization

Parametric 3D modeling is one of those buzz phrases you'll see everywhere in the CAD world. You just saw a simple example of parameterization in the previous recipe, when some details such as width and length were pulled out of the square() command to be stored in their own variables.

Let's consider a more illustrative example. What if you needed to create four rectangles, each one 20 percent bigger than the last? You could do all the math yourself and type in a whole bunch of calculated numbers for widths and lengths, or you could parameterize the situation and make life so much easier.

Here's a set of four rectangles, each 20 percent bigger than the previous one. Each rectangle is also translated out along the x-axis so the rectangles don't overlap and are easier to study. In this first case we'll do the math manually:

```
translate([5,0,0])
square([3,4]);

translate([10,0,0])
square([3.6,4.8]);

translate([15,0,0])
square([4.32,5.76]);

translate([20,0,0])
square([5.184,6.912]);
```

Now let's do some simple parameterization to make life a little easier:

```
size1=[3,4];
size2=size1*1.2;
size3=size2*1.2;
size4=size3*1.2;

translate([5,0,0])
square(size1);

translate([10,0,0])
square(size2);

translate([15,0,0])
square(size3);

translate([20,0,0])
square(size4);
```

This set of commands creates exactly the same set of rectangles as shown above. Each size variable is multiplied by 1.2 to create the next.

But perhaps you change your mind and need the rectangles to be half as wide as they are long, so your first rectangle will be sized [2,4]. In the first case you have to repeat all that repetitive math to resize all the other rectangles. In the second case, you have only to change a "3" to a "2" in one spot, and the change migrates to them all:

```
size1=[2,4];
size2=size1*1.2;
size3=size2*1.2;
size4=size3*1.2;

translate([5,0,0])
square(size1);

translate([10,0,0])
square(size2);

translate([15,0,0])
square(size3);

translate([20,0,0])
square(size4);
```

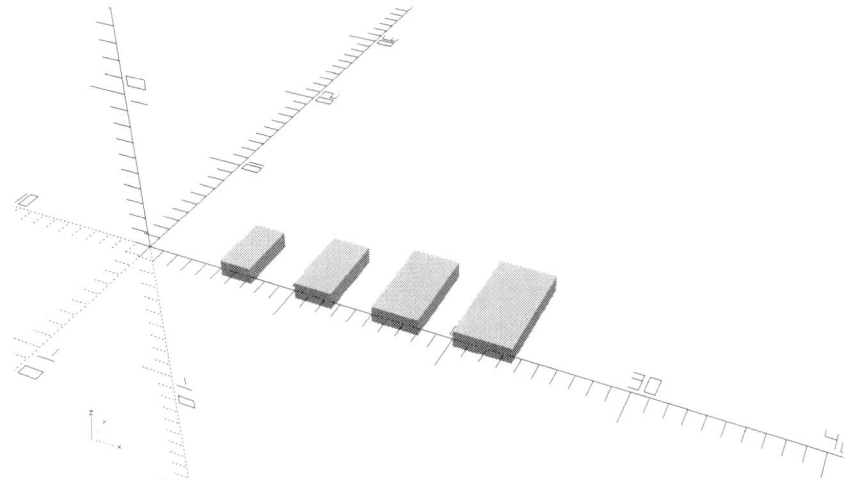

This example is quite simple, but the concept is sound. When you create larger, more complex objects in 3 dimensions, parameterization in OpenSCAD will save you lots and lots of hassles.

The trick is to think ahead, and don't hesitate to put real numbers inside of variables, and even put previous variables inside of calculations to create other variables. Let the variables do the math so you can relax and concentrate on the bigger picture.

Notes

Recipe 4

Create a Circle

In the first recipe we created a sphere with a given radius or diameter, centred at the origin. A 2D circle is created in much the same way, with a given radius or diameter, exactly like pouring a fixed amount of batter on the griddle to create a pancake of a repeatable size, only different. Like the square() command, the circle() command creates a mathematically flat pancaked circle that is rendered on the screen with a thickness of 1 unit. Type this command to create an example circle:

```
circle(10);
```

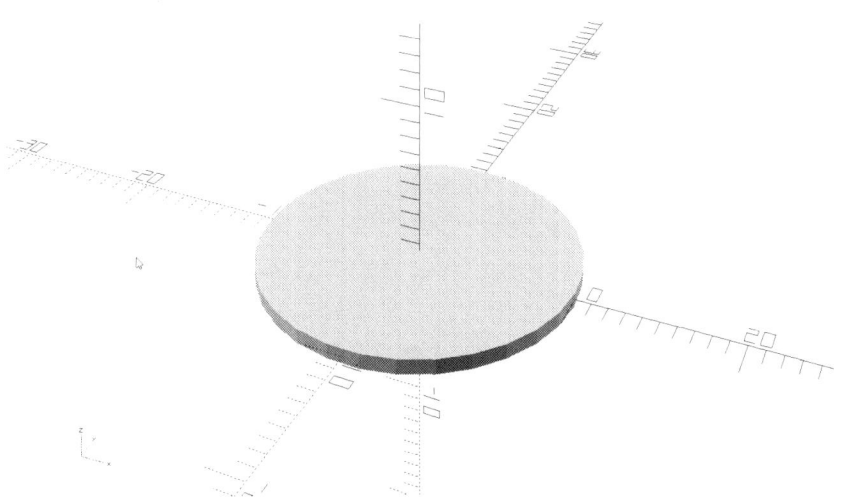

As mentioned before, the magic system variable $fn can be set to change the "number of fragments" for curved surfaces or curved edges.

Keep in mind that in OpenSCAD all edges and surfaces are actually flat or straight, and it's only by breaking those edges up into a significant number of fragments that the illusion of smooth surfaces or edges is created. Let's try a larger value for $fn to smooth the edges of our pancake, I mean circle, just for kicks:

```
$fn=100;
circle(10);
```

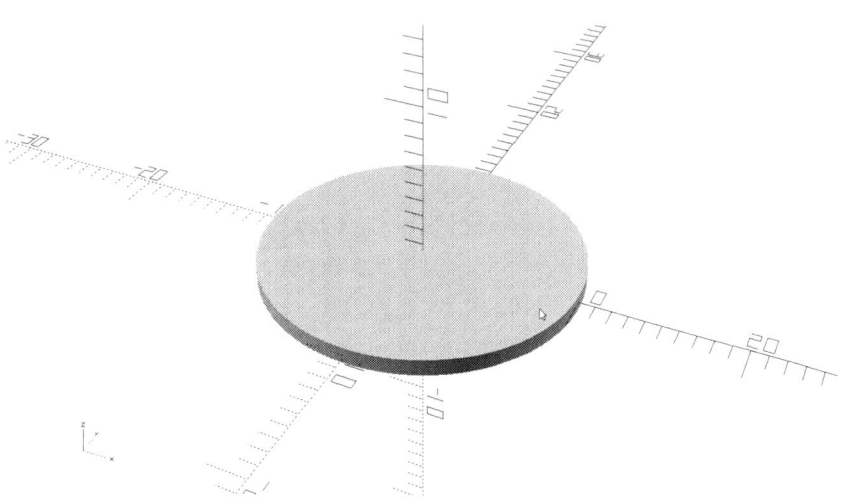

You might be tempted to create an equilateral triangle, a square, a pentagon, a hexagon, and even a heptagon by setting $fn to 3, 4, 5, 6 or 7 respectively. I suggest you give this a try, as it's interesting to see how it works, but do be aware that the OpenSCAD manual strongly suggests not to create modeled objects using this technique, because there's no guarantee that any graphics hardware or software will follow this suggestion and instead create a smooth circle if it knows how. There are better ways to create regular polygons, to be covered later.

As one example, here's a pseudo-pentagon shaped pancake created as a circle where the number of fragments has been set to 5:

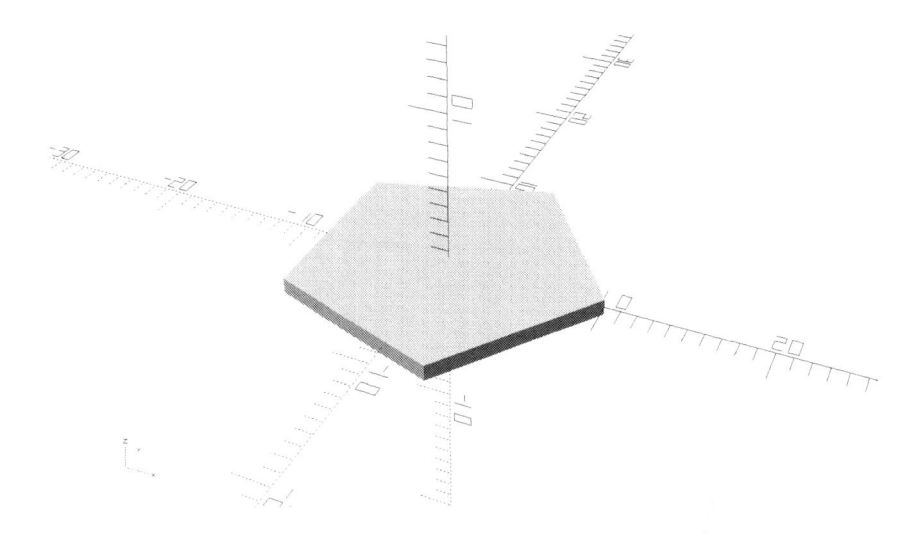

```
$fn=5;
circle(10);
```

Notes

Notes

OpenSCAD Cookbook

Recipe 5

Rotation and Translation

So far we've just created simple geometric objects in the form of flat, 2D slabs lying in the x,y plane. There are two modifying commands that are so important that this recipe is devoted to experimenting with them.

The rotate and translate modifiers act in a very specific order and manner to move and spin objects in space before being rendered. Rotate and translate work on all primitive drawing objects, and even grouped objects, so what is demonstrated here applies to just about everything you'll be creating. I like to picture these commands like flipping, spinning, and generally tossing about the kitchen my pancake creations. Get the hang of how they work and you'll be well on your way to becoming a 3D modeling chef/guru!

Type in the following two command lines, see what they create, and then we'll discuss the details:

```
rotate([0,45,0])
circle(10);
```

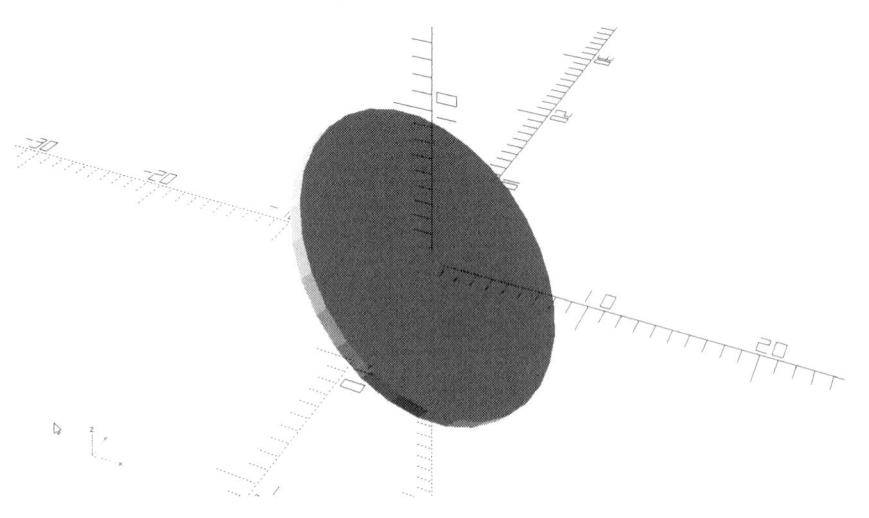

First, notice that the rotate() command does not end with a semicolon. This is because it is a modifier for the circle() command. One or more modifier statements can precede any object-creating action command. Once the object is created, all of its modifiers are forgotten. In other words, consider that all circles are created in the x,y plane centered at the origin, unless each circle is specifically modified using statements such as translate() and rotate(). I like to think of each semicolon as the point at which all modifying statements are reset to zero.

Now let's visit that rotate() statement presented above. The rotate modifier is passed a list (remember the square brackets) of three rotation angles in degrees. Three things happen to our circle here, a rotation about the x-axis, followed by a rotation about the y-axis, and finally around the z-axis. Kind of like flipping a pancake left to right, then top to bottom, and then standing it on end. Or something like that - I think you get the picture. In our example, no rotation happens around the x and z axes, but a 45 degree rotation in space happens around the y-axis.

Right hand rule

Notice the little orientation graphic in the bottom left corner area of your display. This is handy for visualizing the direction of the three axes, especially useful if you've spun the view of your creation manually using your mouse. Take a look to see in our illustration, that the y-axis is running from the bottom left foreground to the top right background. The y-axis rotation has spun the circle around this y-axis such that the circle no longer lies flat in the x,y plane.

Rotations in OpenSCAD follow the right-hand rule. If you point the thumb of your right hand along the positive y-axis, your fingers curl around your thumb in the direction of a positive rotation. I like to imagine firmly grabbing an axis line in my right fist, with my thumb pointing away from the origin. My fingers are then curled around in the direction of a positive rotation.

Study the position of the circle above to see how it has been spun around the positive y-axis by 45 degrees, causing its right edge to dip and its left edge to rise.

Let's try one more rotation command, and then I strongly suggest spending some time experimenting to make sure you understand how the rotations work. Try this one:

```
rotate([0,90,45])
circle(10);
```

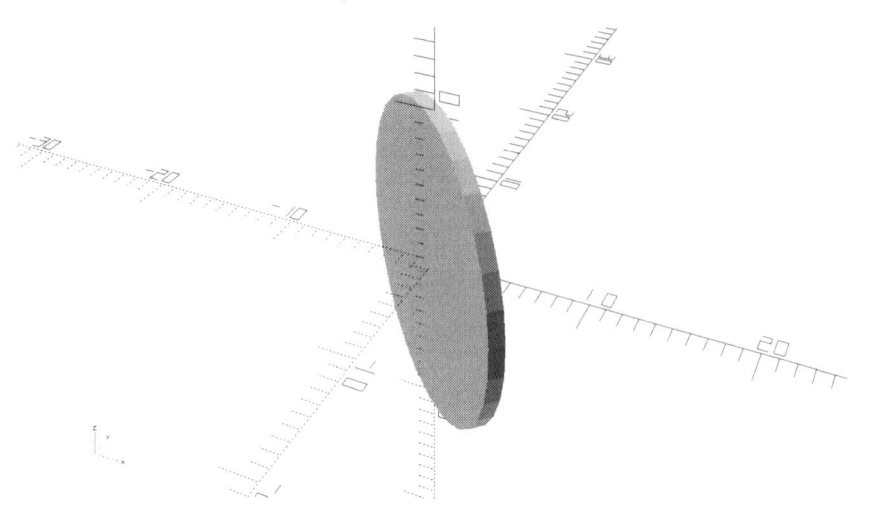

This circle is first spun around the y-axis by 90 degrees, causing it to stand up on edge facing the positive x-axis. Next the 45 degree rotation around the z-axis spins it around the vertical z-axis to the orientation shown. For this second spin the right hand rule has your thumb pointing straight up, with your fingers curled around to spin the circle into the final position shown.

Order is important

The order of the spins around the axes is important. If you want to spin around the axes in an order other than x, then y, then z, consider using more than one rotate command, each statement in order spinning around one axis by setting the other two numbers to zero. For example, the following three statements cause a circle to spin first around the y-axis by 45 degrees, and then by -45 degrees around the x-axis.

This brings up an important point to always keep in mind. Start at the action command's semicolon and work backwards! In the following, a circle is to be created, but before it is rendered on the screen it will be rotated by 45 degrees around y, then it will be rotated around x by -45 degrees. The modifying statements are stacked up and processed in a last-in-first-out order. (This will be easier to see in a minute when we discuss the translate() modifier.)

```
rotate([-45,0,0])
rotate([0,45,0])
circle(10);
```

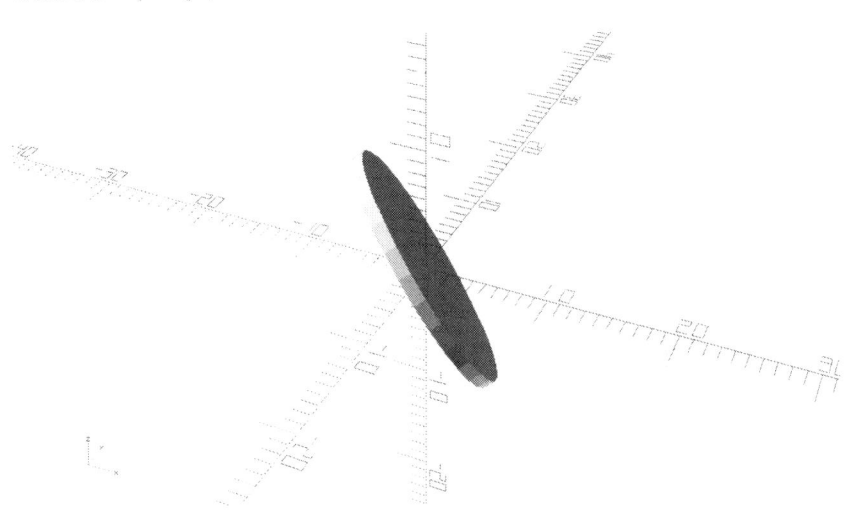

Don't worry if the rotations are tricky to imagine before they happen. The beauty of OpenSCAD is that you can interact with your creations to try various rotations and translations, providing the feedback to help you grasp what's happening in space. This really does help, and over time the predicted rotations and translations become much easier to grasp.

Translations work in a very similar way to rotations. Try this example:

```
translate([10,10,0])
circle(10);
```

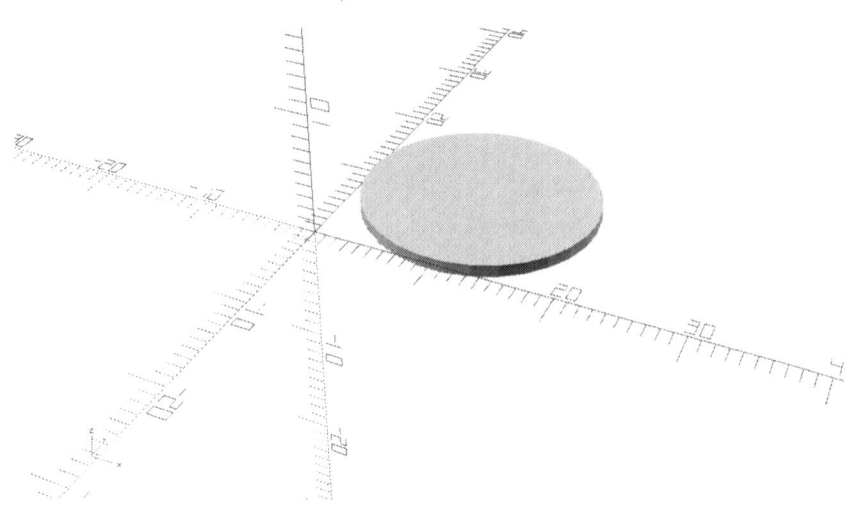

The circle is translated along the positive x-axis by 10 units, and along the positive y-axis by another 10 units. You've slid your pancake to the upper right quadrant of the griddle.

Yes, you can lift the circle up in the air, so to speak, by a translation along the z-axis. Grab a digital spatula and give that a try:

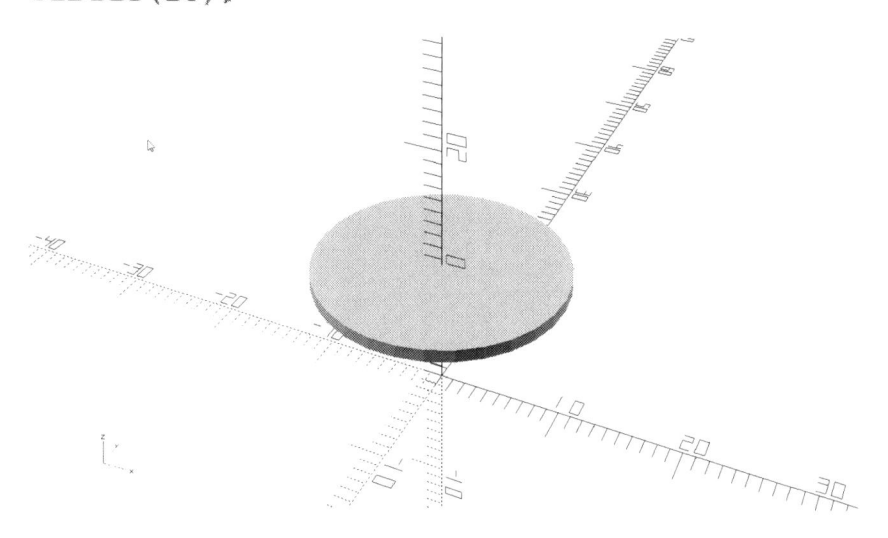

```
translate([0,0,10])
circle(10);
```

It's important to understand the results of combining rotations and translations, especially how the order affects things. The following two circles are modified using identical rotation and translation commands, except that the order for each is reversed.

```
translate([20,0,0])
rotate([0,-30,0])
circle(10);

rotate([0,-30,0])
translate([20,0,0])
circle(10);
```

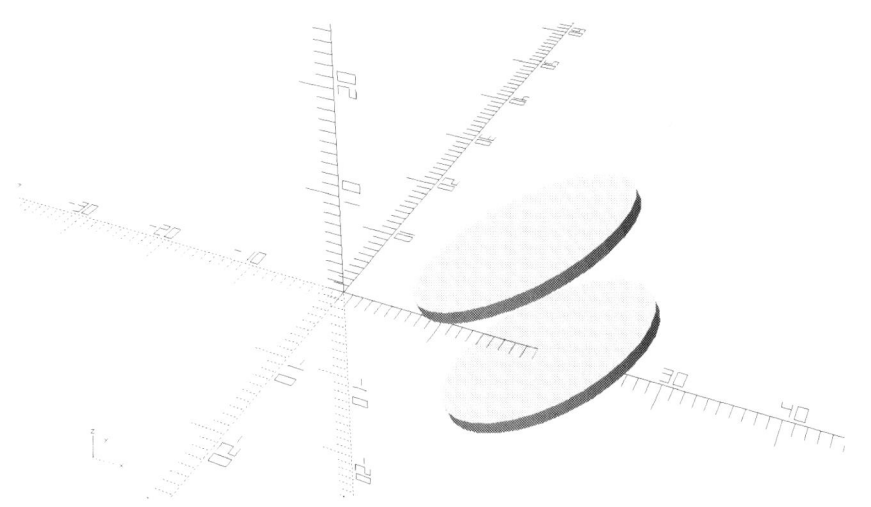

The bottom circle is created by the first set of code. It starts centered at the origin, where it is first rotated -30 degrees around the y-axis to lift its edge, and then translated along the x-axis by 20 units. As a result its center remains on the x-axis.

The top circle is created by the second set of code. It also starts at the origin, where it is first translated out 20 units along the x-axis. Rotations are always around the axes as they pass through the origin, so the whole circle is lifted up as it rotates around the y-axis.

Again, stay clear on the concept that the modifying commands are stacked up and processed in what appears at first to be a reversed order. Picture it like this; the pancakes are stacked up, and once the semicolon is encountered, they are all eaten or processed starting with the most recently stacked pancake and working back until they are all gone.

Notes

Recipe 6

Create a Polygon

In addition to 2-dimensional circles and squares, OpenSCAD lets you create polygons with as many edges as you want - you just provide a list of corner point coordinates.

To create regular polygons, such as equilateral triangles, hexagons, and so on, you'll need to do the math to determine the coordinates, but in a later recipe we'll create a subroutine to effectively expand OpenSCAD and make that process easy. For now, let's create some simple polygons to get the hang of the polygon() command:

```
polygon([[0,0],[30,0],[0,40]]);
```

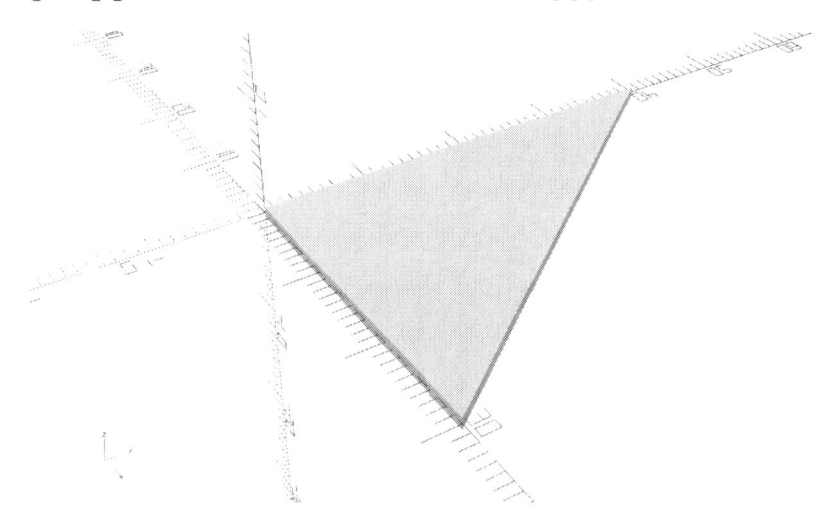

Recall that lists in OpenSCAD are comprised of a sequence of numbers separated by commas and enclosed in square brackets. A 2D coordinate point, representing a point in the x,y plane, is created as a list of two numbers. For example, the point [3,4] is located at x=3, y=4, and z=0. The polygon() command takes a list of these 2-dimensional points and connects them sequentially to form a polygon in the x,y plane.

In the example above, we've created a 30, 40, 50 right triangle by connecting three points, starting at the origin and connecting to [30,0] and [0,40]. Notice that the three coordinates are enclosed in an all encompassing pair of outer square brackets. The polygon() command wants a single list of coordinates, so we pass it a list of points. Technically this is a list of lists, but it's easier to think of it more simply as a list of coordinate points.

The polygon() command can also optionally take a second list comprised of indexes into the first list to define the order to connect the points. I'm sure there are times when this can be useful, but in all of the designs I've created so far I have yet to need it. Keep this detail in mind in case you encounter some obscure design where it would be helpful, but in almost all cases, simply list the coordinates in the right order to "walk around" the shape of your polygon and you'll do just fine.

Here's an example of a more complicated polygon. Note that the points in this case are defined as list variables outside the polygon command, making the code easier to read and modify:

```
p1=[20,20];
p2=[10,25];
p3=[-15,12];
p4=[-22,-5];
p5=[0,-20];
p6=[0,-5];
polygon([p1,p2,p3,p4,p5,p6]);
```

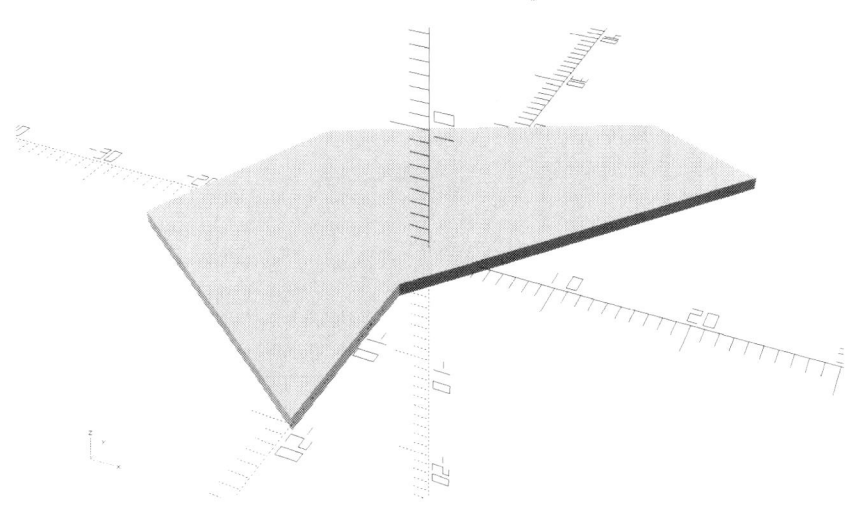

Try changing some of the coordinates or adding some more points to see how you can create the tasty polygon of your dreams.

Of course the translate and rotate modifiers let you move your "flat" (1 unit thick on the screen) polygon around in space anywhere desired. Let's return to the 30,40,50 right triangle and adjust its location in some wild way:

```
// Define the corner points
p1=[0,0];
p2=[30,0];
p3=[0,40];

// Create and position a triangle
translate([30,20,10])
rotate([0,0,180])
polygon([p1,p2,p3]);
```

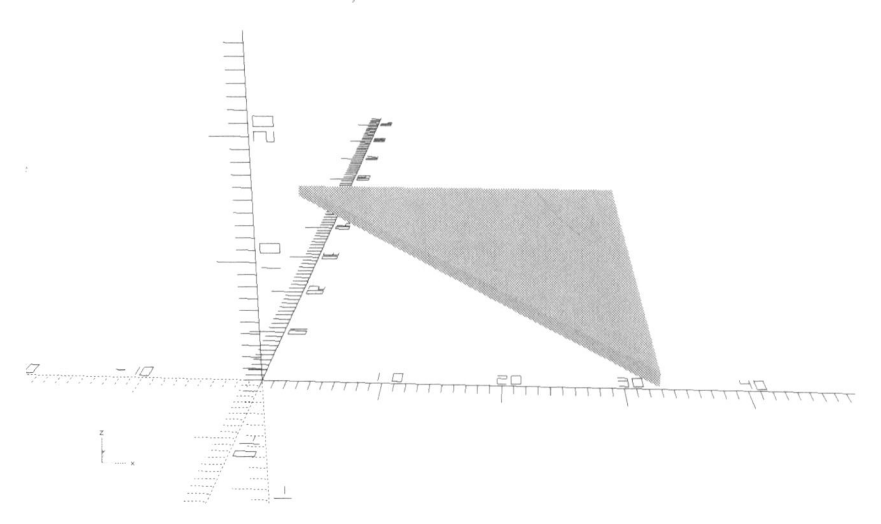

This is the first recipe that shows comment lines - the ones with the double-slashes at the start of the line. Commenting your code is a very good habit, especially when you create code one day and come back to it in three weeks. Comments do help you recall what the heck you did. Comment lines are ignored by the rendering engine in OpenSCAD, so feel free to say almost anything you want in those lines, depending on who might end up reading your code, of course.

In this example a 30,40,50 right triangle is created like the first example in this recipe, but it is then rotated around the z-axis by 180 degrees, then shifted over, up, and out by 30 units along x, 20 units along y, and 10 units along z. (Always remember the "reverse" order of the modifier commands.) The illustration shows the result, but getting a feel for exactly where and how the triangle is floating in space is best gained by rotating, zooming, and panning the results with your mouse. Give it a try.

Notes

Recipe 7

Trimming the Edges

All of the 2D shapes we've created so far have sharp points and straight edges, except for the circle of course. The offset() command provides a way to round or chamfer the corners while expanding the edges out further, or to pull them in a bit and do the same.

To demonstrate how the offset() command works, let's create a clever cleaver by combining three "square" rectangles to form a 2D shape that looks a little like a cleaver. Yes, it needs a little fleshing out, so to speak, but it will suffice for now to demonstrate the offset command.

```
// Clever Cleaver

//offset(r=5)
//offset(delta=5)
//offset(delta=5,chamfer=true)

rotate(10)
translate([100,25,0])
{
   square([100,3]);

   translate([50,0,0])
   square([70,25]);

   translate([0,-3,0])
   square([5,9]);
}
```

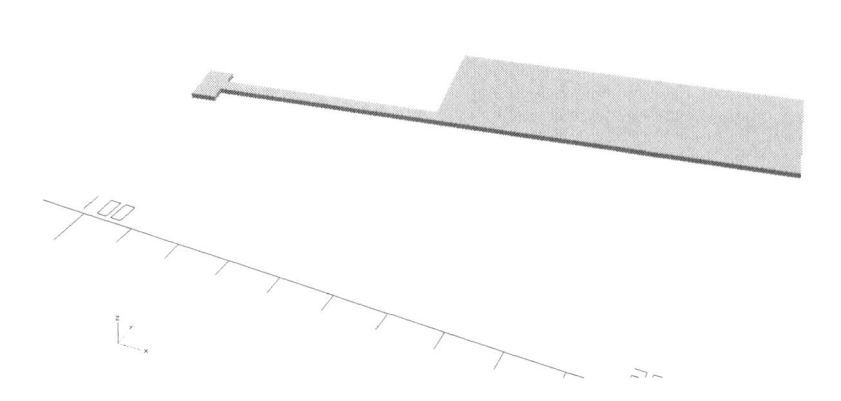

To start with, all of the offset commands were commented out in this first example, so we can see the unaffected outline of the cleaver. Remove the comment slashes to try out the first offset command:

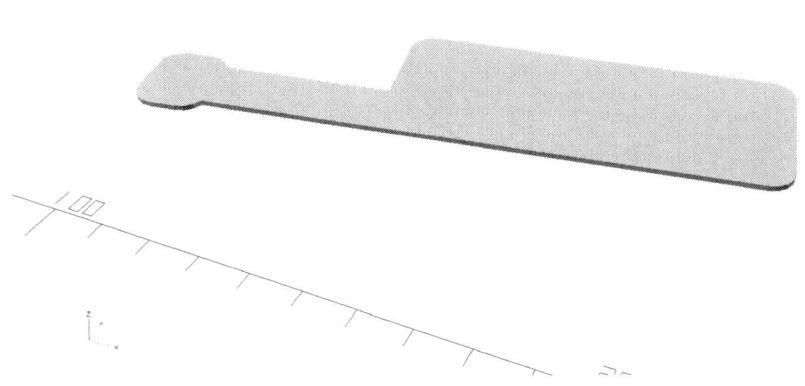

```
offset(r=5)
```

All edges are expanded out by 5 units in the direction away from the interior of the shape, and all corners are rounded as sections of circles with a radius of 5 units. The result is a smoother appearance, with no sharp points. As a clever cleaver you'll need to sharpen the cutting edge, but that's another story.

Remove the comment slashes one at a time from the other offset commands to see how they work. For example, the value of delta shifts the edges out by the indicated number of units, but no rounding occurs:

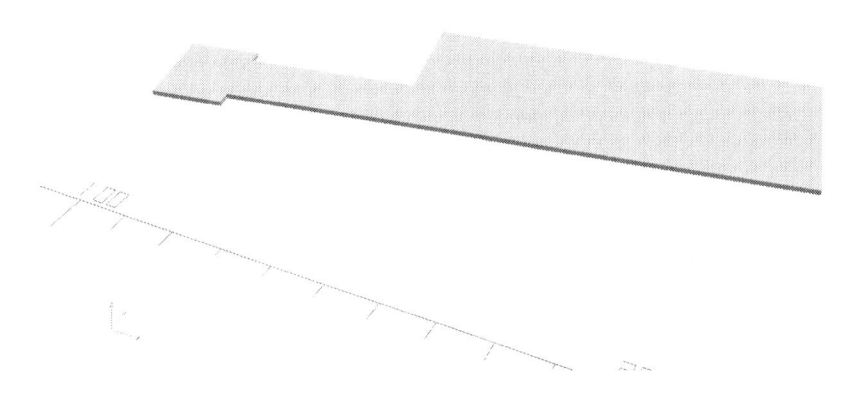

```
offset(delta=5)
```

The optional chamfer parameter, when set to true (it defaults to false) causes the delta corners to be cut off with one flat cut:

```
offset(delta=5,chamfer=true)
```

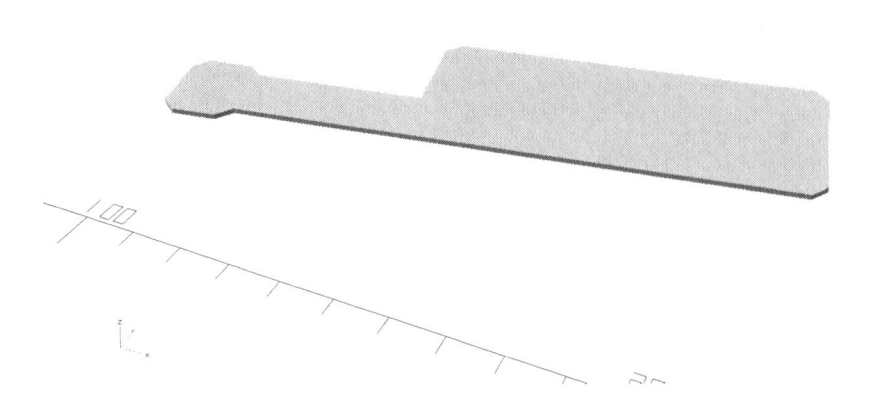

The chamfer parameter only works with the delta parameter, and has no effect if r is set.

If you set the value of either r or delta to a negative value, the edges will move inwards, towards the interior of the shape. Try using -1 for r and delta in the above offset commands to experiment with the cleaver shape.

Note that the offset() command works only with 2D shapes. Later on we'll use the Minkowski() command to do something similar as we round the corners on a 3D shape.

Notes

Recipe 8

Stamp Your Name On It

Once you have your invention ready to print, or perhaps ready to have a professional steel or aluminum mold made for mass production, you might want to put "Patent Pending", or perhaps your company name or other information somewhere on it. Or perhaps you've created a battery holder slot and simply want to emboss a big "+" sign near where the positive pole of the battery should be inserted. Fortunately, OpenSCAD now lets you add text to your objects, like putting frosting and "Happy Birthday!" on your cake.

Text in OpenSCAD is a 2D object, just like the square, circle, and polygon shapes in the previous recipes. Don't worry, this is the last of the 2D items, and starting with the next recipe we'll start extruding these, and start working directly with real 3D objects. For now though, let's create some one-unit-thick (on the screen) text to get the hang of the text() command:

```
text("Happy Birthday!");
```

That was easy enough! The text command does have a lot of optional parameters to control many details of the text, such as size, font, alignment, spacing between characters, and so on. You'll want to refer to these details pretty much every time you create some text. The best way to do this is to find the text command in the official Cheat Sheet at http://www.openscad.org/cheatsheet and click to go to the page that provides all these details.

Smoothness

For now, let's discuss a few details that you should keep in mind. For example, the characters in the text are created behind the scenes with many of the same algorithms as for other primitive objects. In particular, the $fn system variable determines how smooth the curved parts are rendered. Zoom in on the letters in our example to see the flat fragments before and after setting the value of $fn:

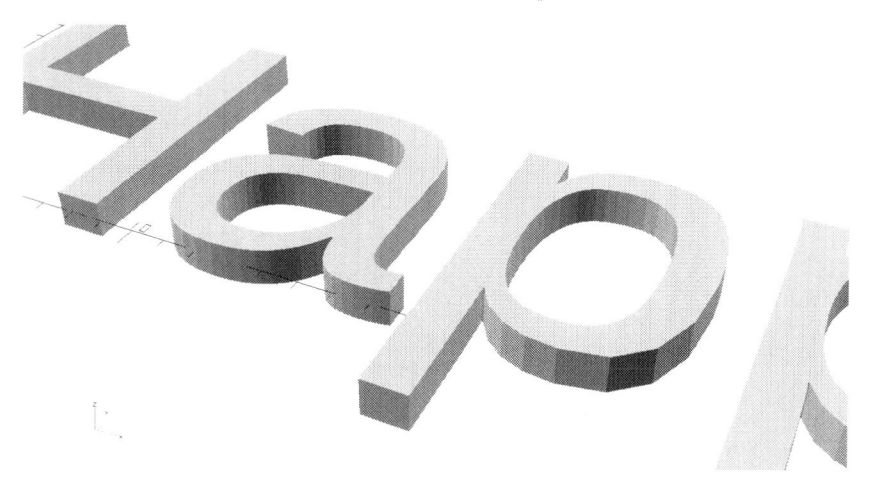

Compare the rough smoothness of the curved faces above, using the default rendering of our example text, with the following creamier smoothness using a bigger value for $fn:

```
$fn=100;
text("Happy Birthday!");
```

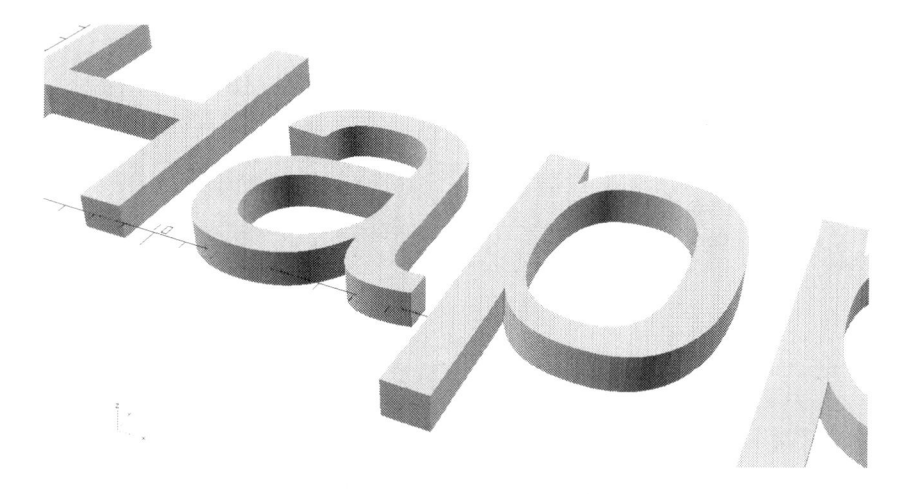

As demonstrated here, the text characters look like they would create a whole bunch of separate objects if sent to a 3D printer. For real projects you should add text onto other objects, to combine them into a single object that either has text sticking out from its surface, or recessed to form hollowed out characters. In the next recipe, you'll plop this text onto the face of a box-like object, using the linear-extrude() command to make it all work.

Notes

Recipe 9

Extruding Into Space

The previous recipes have created 2D objects that lie in the x,y plane. They are all rendered in preview mode on the screen with a thickness of 1 unit, but mathematically they have no thickness in the z direction. If you press F6, or click on the Render icon in the toolbar, the preview becomes a full CSG render (Constructive Solid Geometry), and at that point you'll see the object's thickness go away. If, at this point, you try to output an STL file, it will fail with an error message stating the object "is not a 3D object". Makes sense.

The next step is to give these 2D objects some real thickness by using the linear_extrude() modifier command. There's a lot going on with linear_extrude, so let's break it down and experiment with its various parameters to see what happens. Start with a simple example that extrudes a flat square into a true cube 10 units on all edges:

```
linear_extrude(10)
square(10);
```

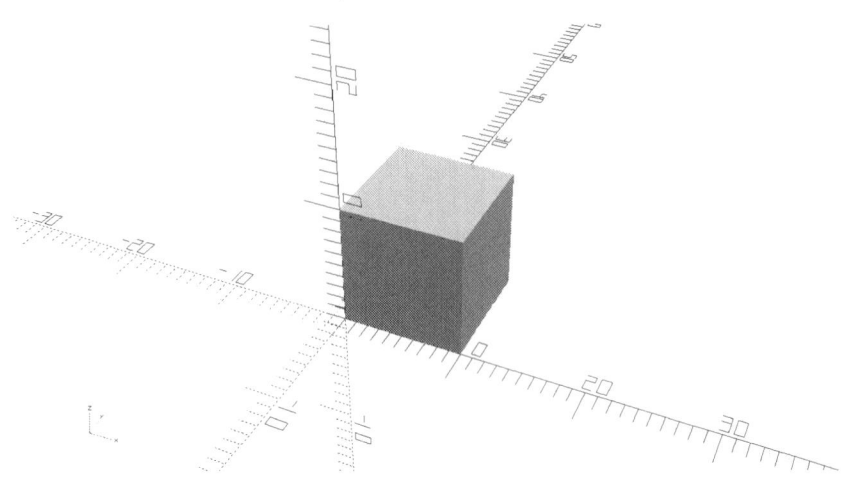

The extrusion can be very short or very long. For example, here's a long rod shape created as a circle with radius 10 that is extruded 200 units in the y direction. Perhaps this is the start of a birthday candle for the top of your cake:

```
linear_extrude(200)
circle(10);
```

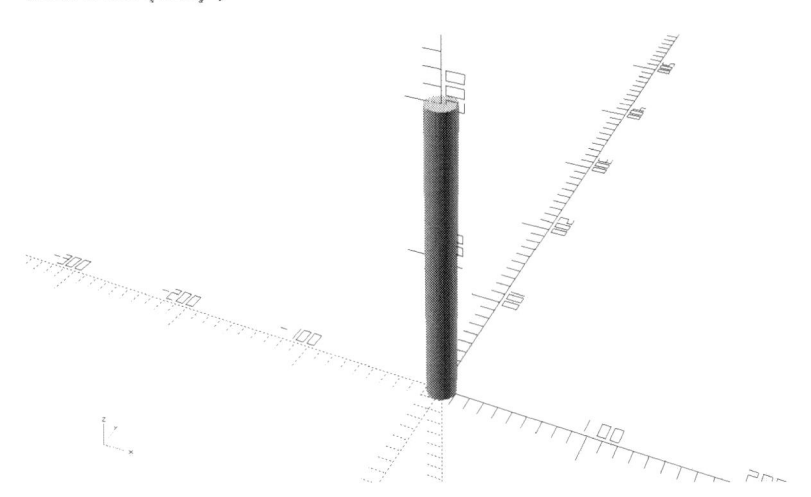

As mentioned, there are a lot of powerful parameters to control the linear_extrude command. One of the simplest is the addition of "center=true", which causes the extruded shape, whatever it may be, to be centered half above the x,y plane and half below. Here's a square, size 10, that is extruded to a height of 30 units, but centered such that the top is at 15 and the bottom is at -15 units:

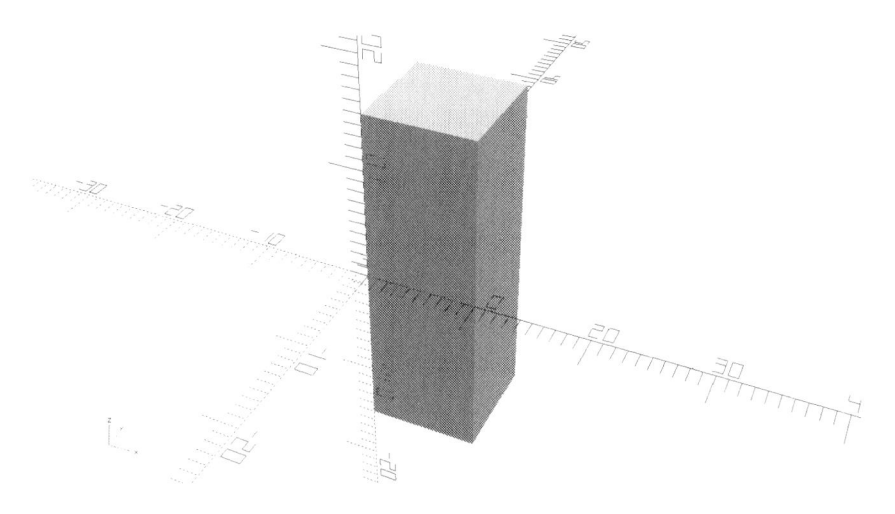

```
linear_extrude(30,center=true)
square(10);
```

A really wicked and twisted parameter (in a good way) is twist, which literally adds a twist to the linear_extrude command. There are lots of subtleties here, and the best way to grasp them is to give them a try.

Verbally described, the twist parameter causes the 2D object's projection on the x,y plane to be rotated around the z-axis by the given number of degrees from bottom to top of the extrusion, using the left hand rule. Whew! Like I said, a few examples will help clarify, and once you get the hang of how the twist works you'll see it adds some powerful twists to the linear_extrude command!

Start with a centered square that is extruded 5 times is size and with 180 degrees of twist:

```
linear_extrude(50,twist=180)
square(10,center=true);
```

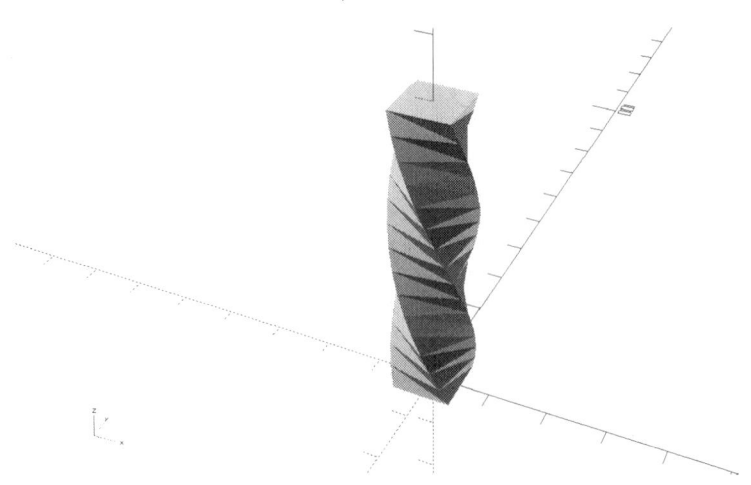

This twist rotates the square around the z-axis by 180 degrees during the extrude to form a solid object. If you follow the edges you can see that the front edge of the square is rotated to the back from bottom to top.

The slices parameter is kind of like the $fn parameter in that it smooths things out when the going gets rough during a linear_extrude. Let's repeat the same extrusion of the square, but with slices set to 100:

```
linear_extrude(50,twist=180,slices=100)
square(10,center=true);
```

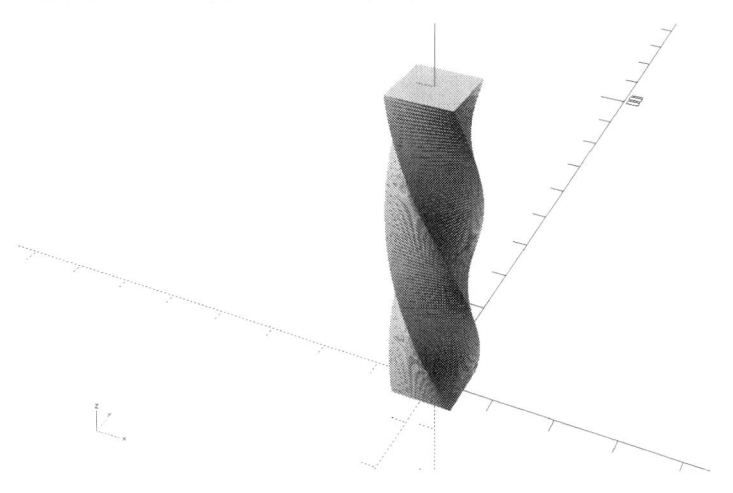

To clearly demonstrate how the twist always happens around the z-axis, translate the square away from the z-axis before the extrude. Also, to make the twisting more dramatic let's increase the total twist rotation angle to 720 degrees, or fully twice around the z-axis. We're creating a curly fry in this recipe, as you'll see:

```
linear_extrude(50,twist=720)
translate([10,0,0])
square(10,center=true);
```

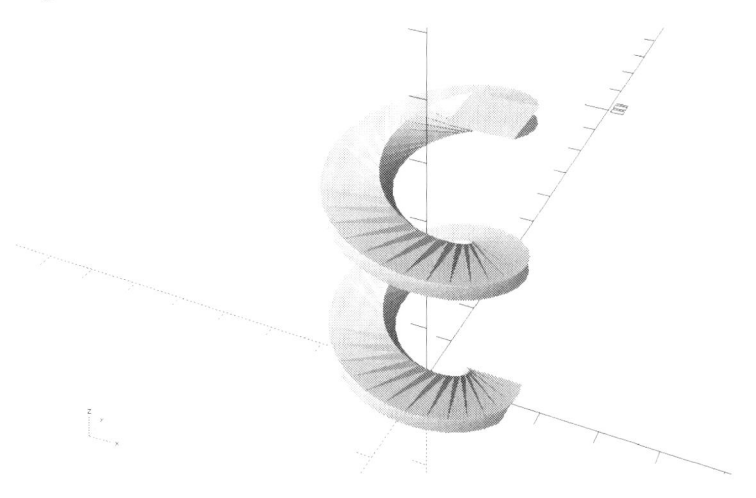

Of course the twist works with the extrusion of any 2D object, or more accurately, with the projection onto the x,y plane of any object. Consider a circle that has been flipped up a bit, shifted out of the x,y plane, and translated away from the z-axis. The following creates a circle twice, where one copy is placed in space as described, and the second copy is extruded from there to create a strange but dramatic spiral.

The circle's projection (visualize it's "shadow" when the sun is straight overhead) on the x,y plane is an ellipse, because of its 60 degree tilt. The extrusion then rotates the ellipse starting on the x,y plane, rotating around the z-axis by 720 degrees.

By far the best way to get a feel for this one is to use your mouse to spin the results around to view it from many angles, and to try changing the various numbers involved. Do give it a try, as it demonstrates most of the tricky details of extruding with a twist:

```
angle=60;
shift=25;
size=10;

translate([shift,0,shift/2])
rotate([0,angle,0])
circle(size);

linear_extrude(50,twist=720)
translate([shift,0,shift])
rotate([0,angle,0])
circle(size);
```

Another powerful but rather obscure parameter is scale, which can be set with either a single number or a two-number list in the linear_extrude command. Later in this book, the stand-alone scale() command is shown to take a list of 3 numbers, but here as a parameter for the linear_extrude() command, it works only in two dimensions.

The linear_extrude starts out on the x,y plane with no change due to the scale parameter, but by the time the extrusion reaches the top the full effect of the scale is in effect. Halfway up, the scale multiplier is halfway in effect, and so on.

The scale parameter can be combined with twist and other complexities for very creative results, but here let's use it on its own for clarity, to create a dramatic base stand for your plate of cookies perhaps:

```
linear_extrude(10,scale=.5)
square(10,center=true);
```

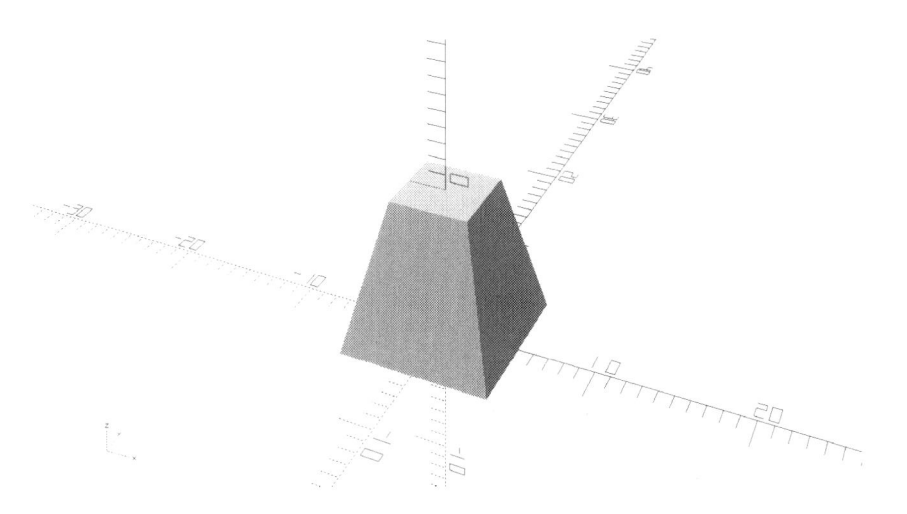

The square is created in the x,y plane at 10 units on each edge, but by the time the square is extruded to a height of 10 units, the scale parameter of ½ (.5) causes the shape to taper to a square of 5 units.

Next try changing scale to the two-numbered list [0,1]. This causes the square to not taper at all in the y direction, but completely taper to zero in the x. The result is a wedge shape, where the "square" at the top has a width of zero units, and the other two edges merge to form a single ridge line, still at 10 units:

```
linear_extrude(10,scale=[0,1])
square(10,center=true);
```

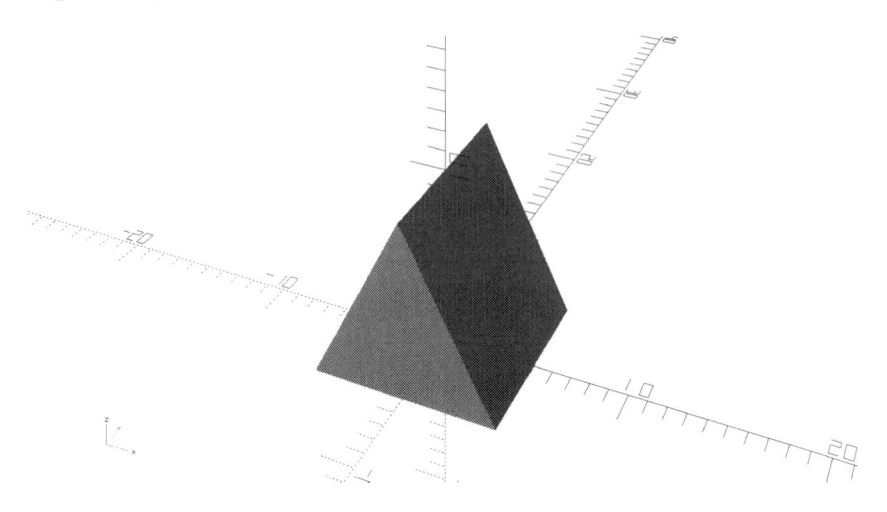

Set the scale parameter to zero to create a pyramid with base edges and an overall height of 10 units each. The "square" at the top has shrunk to a point:

```
linear_extrude(10,scale=0)
square(10,center=true);
```

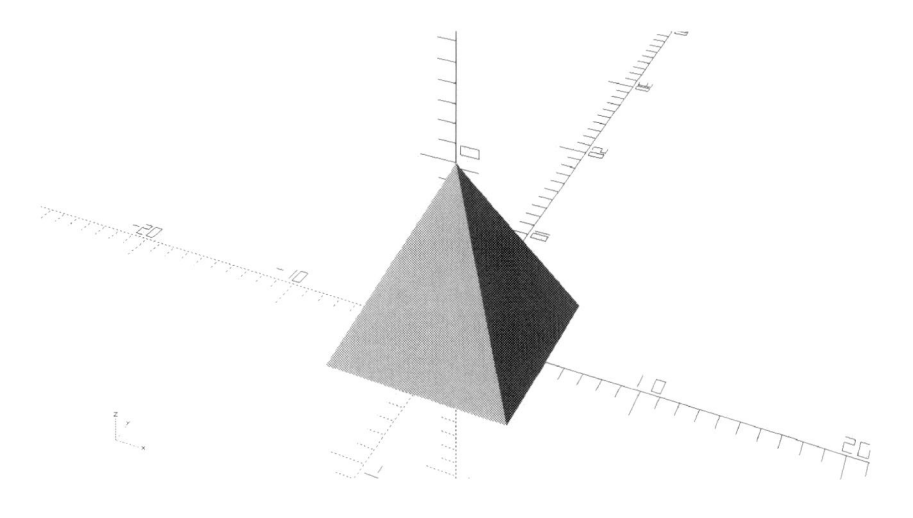

Finally, take away the centering parameter, so the base square has one corner at the origin, and study how this affects the pyramid:

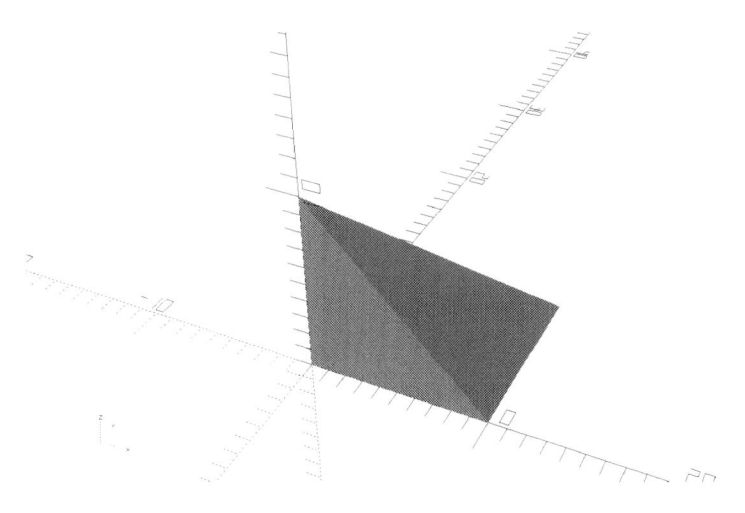

```
linear_extrude(10,scale=0)
square(10);
```

The scaling is applied numerically to all coordinates as the extrude goes up, so the corner coordinate at [10,0] shrinks to [0,0]. All points are effectively scaled towards or away from the origin.

Notes

Notes

Recipe 10

Create a Doughnut

In addition to the linear_extrude command, there's another rather bizarre way to bring a 2D, flat object into the 3D universe inside OpenSCAD. It's called rotate_extrude, and it's pretty amazing.

Rotate_extrude takes the projection on the x,y plane of a 2D object, rotates it 90 degrees around the x-axis, then rotates around the z-axis to sweep out a volume. It sounds complicated. And, well, it is.

It's much easier to experiment with rotate_extrude until it suddenly makes sense. A few examples help immensely. Let's start by making a doughnut, by using rotate_extrude on a flat circle:

```
color("chocolate")
rotate_extrude()
translate([10,0,0])
circle(5);
```

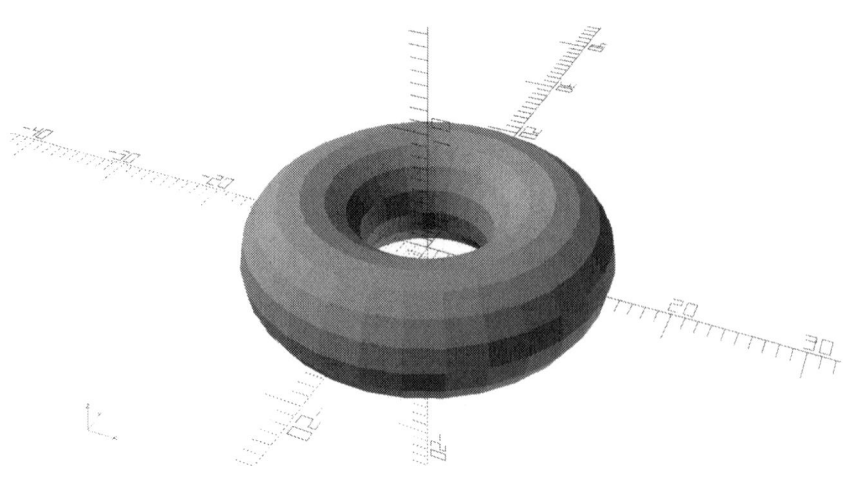

The circle is created in the x,y plane, then shifted out along x to get it away from the origin. The rotate_extrude command then flips the circle up on edge and sweeps around the z-axis to create the doughnut.

This is the first recipe that uses the color modifier. You can color objects using a long list of color names, including chocolate (I thought that was appropriate for our doughnut), or you can build your own color using red, green, blue, and alpha values in a list. Unlike most other programming languages, OpenSCAD chose to define these color parameters as floating point numbers between 0.0 and 1.0, instead of using integers between 0 and 255. Alpha controls the transparency, with 0.0 being invisible, and 1.0 being completely opaque.

For example, color([.5,.5,1]) will turn our chocolate doughnut into a blueberry delight, and color([.5,.5,1,.2]) fades it to appear like the ghost of blueberries past.

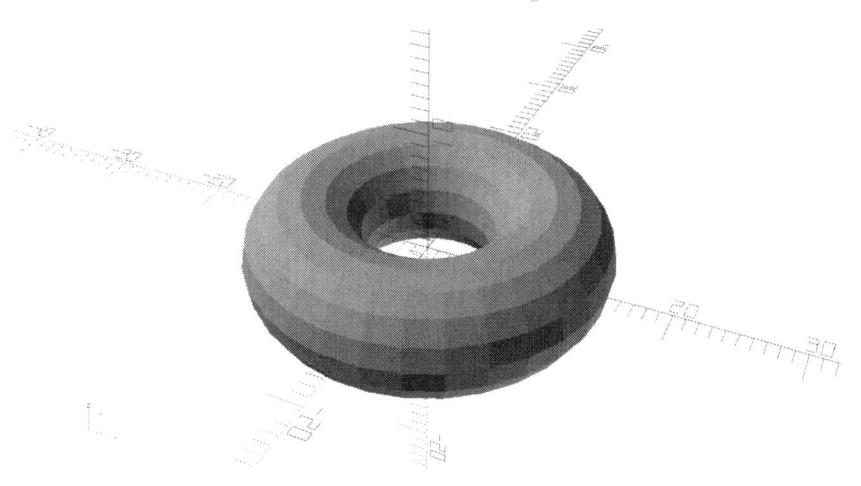

Note that it is important for the entire 2D shape to be in either the positive or negative x half of the x,y plane before applying rotate_extrude. Don't straddle the y-axis! You'll get an error message if you do.

Here's a slightly more useful example of rotate_extrude. The outline of a slice through half of a wine glass, chalice, or perhaps the holy grail, is sketched in the x,y plane as a polygon:

```
thick=1.5;
points=[
[0,thick],
[10,0],
[10,thick],
[thick,thick+thick],
[thick,20],
[12,30],
[15,45],
[15-thick,45],
[12-thick,31],
[0,21]
];

//rotate_extrude()
polygon(points);
```

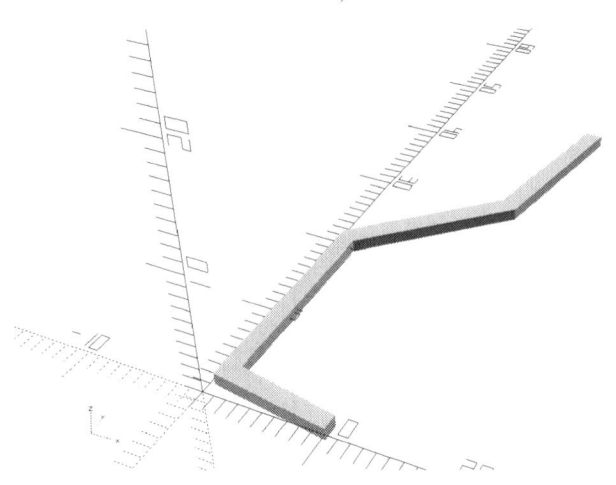

Un-comment the rotate_extrude() command to cause this figure to pivot up to a vertical orientation and then rotate around the z-axis to complete the holy grail:

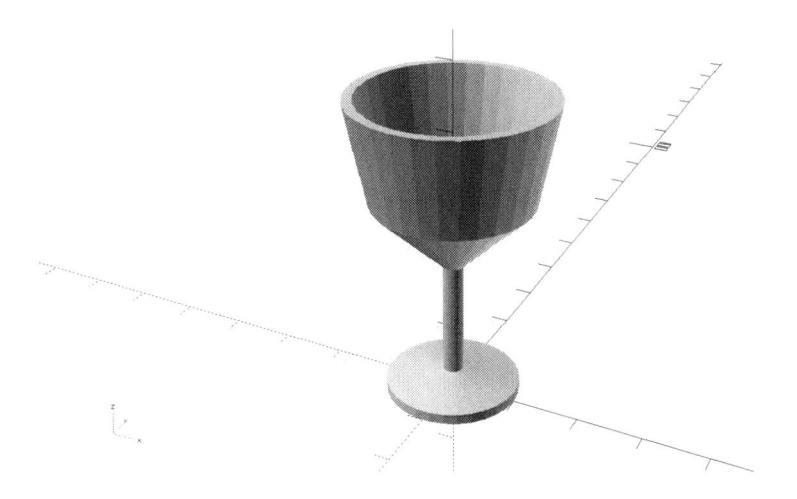

We could punch some holes through this object to create a holey grail, using the difference() function, but I'm jumping the gun a bit, and we'll wait to cover that amazing operator until a future recipe.

A new kind of meatball

OpenSCAD provides enough functionality to accomplish some very creative and remarkable constructions. As a last example in this recipe, let's create a meatball sphere in a completely new way, without using the sphere command. To do this, create a circle, cut away the half in the negative x part of the x,y plane, then use rotate_extrude() to complete the job.

First, here's the half circle, without the rotate_extrude() command:

```
difference() {
   circle(10);
   translate([-10,0,0])
   square(20,center=true);
}
```

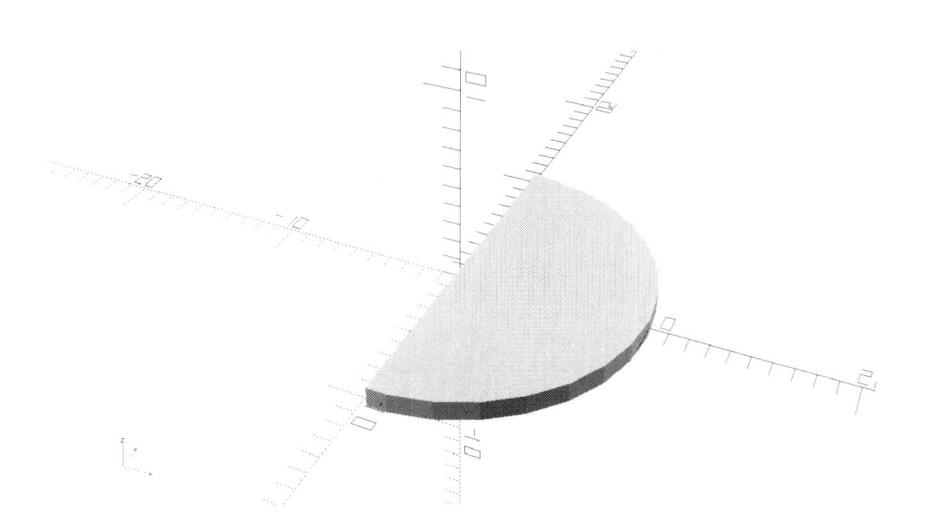

Don't worry about the difference() operator yet, recipe 14 will explain it in more detail. Just be aware that here it is used to remove the left half of the circle by subtracting a square that overlaps the left half of the circle.

Add the rotate_extrude() command to complete the spherical meatball:

```
rotate_extrude()
difference() {
   circle(10);
   translate([-10,0,0])
   square(20,center=true);
}
```

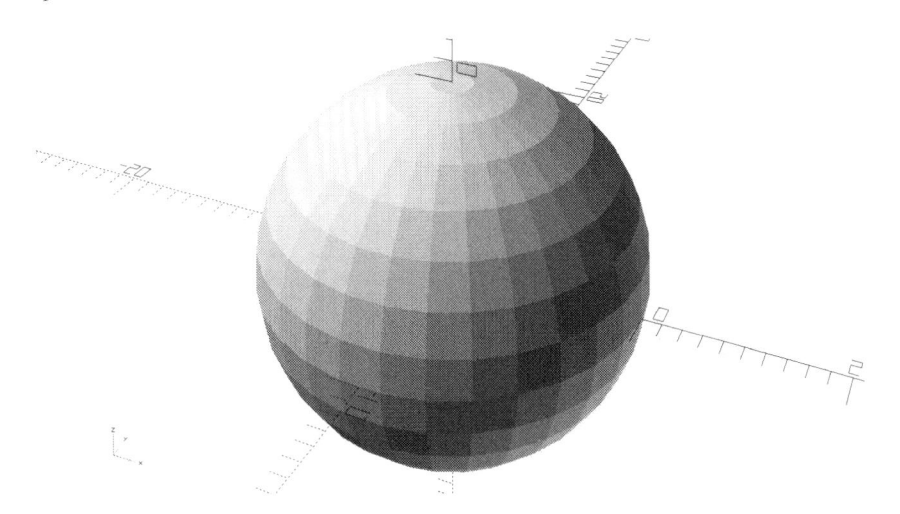

Notes

OpenSCAD Cookbook

Recipe 11

Kitchen Tips and Tricks

While working on a recipe for creating regular polygons, several "gotchas" popped up to cause me hassles. I want to share these with you, because these are details you really need to be aware of, but they are easy to overlook or forget at a later date (my situation). Let's work through a few of them now.

Start with this code listing, which doesn't create a viewable object, but does result in output in the Console window below the main view panel:

```
list = [3,4,5,6,7];
echo(list);
```

```
Console
Compiling design (CSG Tree generation)...
ECHO: [3, 4, 5, 6, 7]
Compiling design (CSG Products generation)...
Geometries in cache: 69
Geometry cache size in bytes: 5763616
CGAL Polyhedrons in cache: 0
CGAL cache size in bytes: 0
Compiling design (CSG Products normalization)...
Normalized CSG tree has 0 elements
Compile and preview finished.
Total rendering time: 0 hours, 0 minutes, 0 seconds
```

The echo command is very handy for displaying intermediate results in your project - an indispensable tool for debugging. Notice the second line in the Console output, where the echo() command has printed the contents of the list variable. The results here are exactly as expected.

You can do simple math on whole lists in one shot. The following code was my first attempt (before remembering a "gotcha") to double all the values in the list above:

```
list = [3,4,5,6,7];
list = 2 * list;
echo(list);
```

Console
Compiling design (CSG Tree generation)...
WARNING: Ignoring unknown variable 'list'.
ECHO: undef

At first, I was mystified. Why in the world would OpenSCAD think my variable named list is now unknown? The answer is tricky, especially if you come from a computer programming background like I do, but it's critical that you understand how OpenSCAD works, and this really is a key point.

Variables are created and populated during source code parsing by OpenSCAD one time, and one time only. The simple rule to remember is to assign a value only once in your program. In this case, the second line is trying to pull in the value of list and then reassign the calculated result to itself. Variables in OpenSCAD are more akin to constants in other languages. As far as the compiler is concerned, list doesn't exist until the last assignment, so the second line of code fails.

The immediate workaround is to create two list variables:

```
list_a = [3,4,5,6,7];
list_b = 2*list_a;
echo(list_a,list_b);
```

```
Console
Compiling design (CSG Tree generation)...
ECHO: [3, 4, 5, 6, 7], [6, 8, 10, 12, 14]
```

Here's an even simpler case that demonstrates this issue. In this case the value of x is set to 200 during code parsing, and 200 is what you, perhaps unexpectedly, get with both echo commands:

```
x=100;
echo(x);
x=200;
echo(x);
```

```
Console
Compiling design (CSG Tree generation)...
ECHO: 200
ECHO: 200
```

There is, however, a way to reuse a variable, in a sense. The for-loop processes a variable with each iteration, but not in the way other languages do. With each iteration of the following, the whole block of code is repeated, with the variable x renamed internally each time, something like x1, x2, x3, and so on. Another way to look at this is that the for-loop processes a list, and the next action command or block of code is repeated for each item in the list, where x is just a local variable placeholder for those items:

```
for(x=[0,7,30,45])
translate([x,x,0])
cube(5);
```

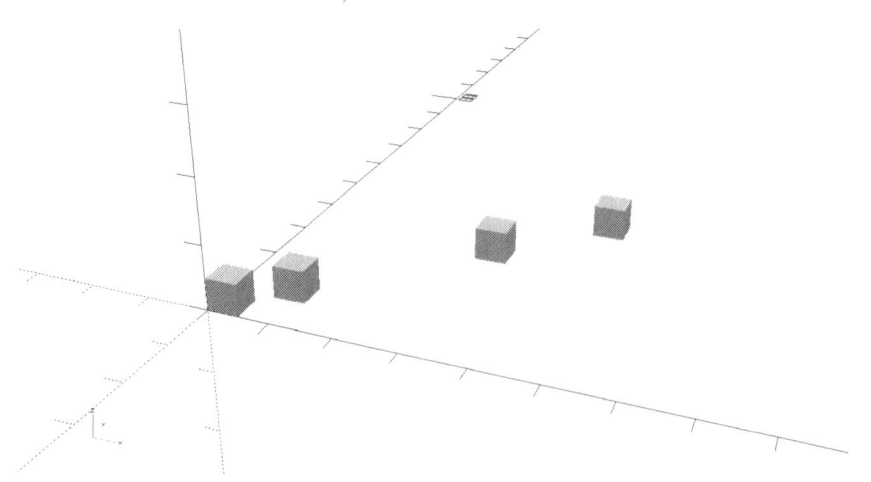

In this example, the for-loop processed items in a list. More often, a range is used to effectively populate a list, and this technique looks more and more like the for-next loops in other languages. Understand the subtle difference and you'll have a good handle on how OpenSCAD's for-loop works.

Ranges

A range looks a little like a list containing two or three items, but instead of commas there are colons separating those items. The range is processed to create a list as instructed by those two or three values. If there are two values then the increment is 1, which is probably the most common way a range is used. If there are three values, the middle one is the increment. Here are a few examples to get you started:

```
for(x=[2:4]) echo(x);        // 2 3 4
for(x=[2:2:5]) echo(x);      // 2 4
for(x=[7:-1:5]) echo(x);     // 7 6 5
for(x=[-1:0.7:1]) echo(x);   // -1 -0.3 0.4
```

The list that the for-loop creates is processed immediately by the next action command or code block by substituting into the variable (x in this case) for each iteration. A slight modification of the code allows assignment of the created list into a variable for later processing. Here's an example that creates a list of square numbers from 0 to 100:

```
sqrs=[ for(x=[0:10]) x*x ];
echo(sqrs);
```

Console
Compiling design (CSG Tree generation)...
ECHO: [0, 1, 4, 9, 16, 25, 36, 49, 64, 81, 100]

For-loops can be nested, but OpenSCAD provides a way to combine them all into one for-loop. Study the following example, and refer to the Cheat Sheet if you need more information on how to use nested for-loops.:

```
for(i=[3,2,4],j=[0:2]) {
   k=i+j;
   echo(i,j,k);
}
```

Console
Compiling design (CSG Tree generation)...
ECHO: 3, 0, 3
ECHO: 3, 1, 4
ECHO: 3, 2, 5
ECHO: 2, 0, 2
ECHO: 2, 1, 3
ECHO: 2, 2, 4
ECHO: 4, 0, 4
ECHO: 4, 1, 5
ECHO: 4, 2, 6

Note that the "i" for-loop iterates through a list, and the "j" for-loop iterates through a range. The local variables i, j, and k are created for each iteration of the entire block of code wrapped in curly braces.

Intersection_for-loop

We'll soon be looking at how OpenSCAD combines shapes using boolean operators such as union or intersection. Don't worry about those yet, but be aware that there's a potential gotcha when creating and combining shapes within a for-loop. What happens is that shapes combined together from each iteration of a for-loop use the union modifier to hook them all together. If you want the intersection of the shapes instead, use the special intersection_for-loop. This seems like a slight kludge to the language design, but it works just fine.

Notes

Recipe 12

Functions, Modules, and Regular Polygons

The goal of this recipe is to create some code to create two-dimensional regular polygons with any desired number of sides. An equilateral triangle, a square, a pentagon, and a hexagon are the first few such figures, with sides of 3, 4, 5, and 6 respectively.

This is the perfect opportunity to introduce functions and modules, OpenSCAD's way to create flexible and reusable code constructs. A function is passed zero or more parameters, and it returns some calculated result, such as a number or a list. A function does not call any action commands in OpenSCAD, such as circle() or square(). A module, on the other hand, is designed to call action commands to create shapes, in addition to performing any required calculations. A module does not return a value.

The following example demonstrates one function and two modules that work together to create n-sided regular polygons.

The function, named cartesian, is passed two parameters. Perhaps you recall, from your calculator days, functions to convert between rectangular and polar coordinates. The cartesian function converts a polar coordinate expressed as a radius from the origin and an angle around (counter-clockwise) from the positive x-axis to an x,y cartesian coordinate location in the x,y plane. As with everywhere else in OpenSCAD, angles are always expressed in degrees.

```
function cartesian(radius,angle)=
    [radius*cos(angle),radius*sin(angle)];
```

The first module is named triangle. This one is really easy, as all it does is receive three point values (each a 2-number list for an x,y point) which it passes on to the polygon command, wrapped in a single list. Basically, this module provides a generic triangle command for readability and understanding, instead of just relying on the polygon command to create triangles.

```
module triangle(p1,p2,p3) {
    polygon([p1,p2,p3]);
}
```

The regular_polygon module calls the cartesian function and the triangle module to simplify the creation of regular polygons of any desired size and number of sides. I've set this module up with default values for the number of sides (3) and radius (1), but normally you'll pass values for both of these parameters to create your desired 2D regular polygon.

```
module regular_polygon(sides=3,radius=1) {
    for(n=[0:sides-1]) {
        p1 = cartesian(radius,n*360/sides);
        p2 = cartesian(radius,(n+1)*360/sides);
        triangle([0,0],p1,p2);
    }
}
```

This module works by looping through all neighboring pairs of points on the perimeter of the polygon, creating a triangle in each case by connecting the two points with the origin, like a slice of pie. As the wedge shaped pieces are created, the default union operation combines them into the completed regular polygon.

Here's sample code to call regular_polygon() to create an octagon with a radius (distance from the origin out to each corner point) of 20 units For a solid result, the octagon is extruded to a thickness of 3 units:

```
poly_edges = 8;
poly_radius = 20;
poly_height = 3;
linear_extrude(poly_height)
regular_polygon(poly_edges,poly_radius);
```

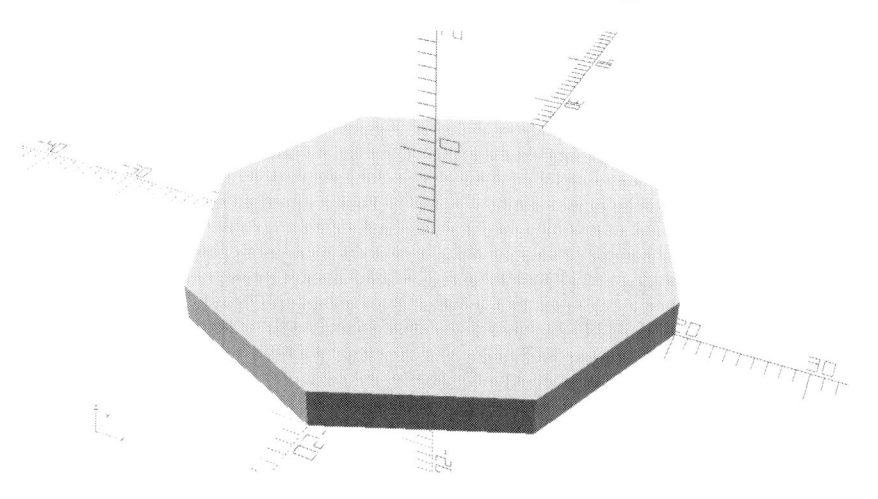

The base shape for my Lucidbrake invention (an octagon shaped automatic brake light for bicycles) and its snap-on polycarbonate lid were created as iconic stop sign shaped octagons in this way.

This would also be a great shape for a "Happy Birthday!" cake for your local school crosswalk stopping guard person thingy. Let's put some red frosting on it, add our "Happy Birthday" plus a smiley face, and make the whole cake a little taller (in the z direction of course):

```
// Parameterized values
poly_edges = 8;
poly_radius = 20;
poly_height = 10;
text_scale = 0.3;
text_translate = [-17,-1.2,poly_height-.1];
text_thick = .4;
font="Liberation Sans";

// The red "stop sign" cake
color("red")
linear_extrude(poly_height)
regular_polygon(poly_edges,poly_radius);

// The verbal frosting on top
color("white")
translate(text_translate)
linear_extrude(text_thick)
scale([text_scale,text_scale,1])
text("Happy Birthday! \u263A",font=font);
```

Okay, you now have your cake, but you can't eat it too. Let's discuss a few of the baking secrets.

The numerical parameters controlling sizes, shapes, and translations of everything are all isolated in a block of code for easy review and maintenance. The first little experimentation challenge is to put the "Happy Birthday!" text up there too, assigned to a variable that is in turn used in the text() action command.

Notice the strange looking characters within the text string. This string ("\u263A")is the Unicode designation for one character, a happy face, which shows up on the cake. There are many other characters available, depending on the choice of font.

Start at the official Cheat Sheet, click on the text() command, and scroll down to learn all about available fonts, Unicode characters, and related information. For this cake, I chose "Liberation Sans" as the font, but there are many others available. The documentation examples show how to add bold, italics, and other modifications to your text.

To make sure the white text intersects with the red cake, the translation of the text in the z direction is slightly less than the height of the cake. Look in the text_translate variable for the z component, "poly_height-.1". By default, OpenSCAD combines shapes that intersect each other in this way into a single shape, using the "union" operator. The next recipe will demonstrate a couple other boolean operators used to combine 3D shapes in powerful ways.

Be sure to play with the various parameters to experiment with this cake. One challenge would be to split "Happy", "Birthday!", and the happy face character onto three lines, to use the surface of the cake more effectively.

Notes

Notes

Recipe 13

No Matter How You Slice It

Engineering drawings, such as in AutoCAD, use 2D projections, or shadows if you will, of 3D objects to quantify their shapes on flat paper. This comes from the old days before digital manipulations of 3D shapes was possible, but there are still times today when these projections come in handy. OpenSCAD provides the projection() command to slice a 3D object to create a 2D shape.

Let's return to our doughnut shape from a couple recipes back, rotated around the y-axis by 45 degrees to make the projection more interesting and informative:

```
rotate([0,45,0])
rotate_extrude()
translate([10,0,0])
circle(5);
```

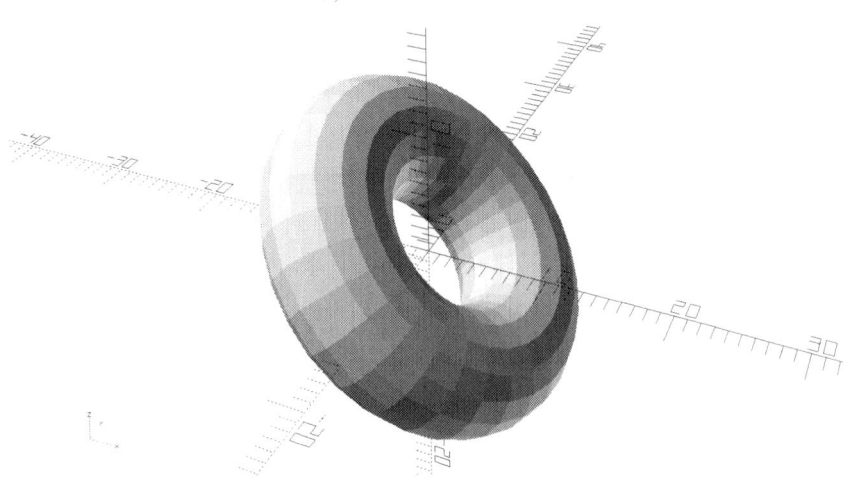

Notice that the doughnut is now straddling the x,y plane, with its left half above and right half below. The center of the doughnut hole is at the origin.

Adding the projection() command reverts the doughnut to a flat shape, where all points on the doughnut are projected down or up to the x,y plane:

```
projection()
rotate([0,45,0])
rotate_extrude()
translate([10,0,0])
circle(5);
```

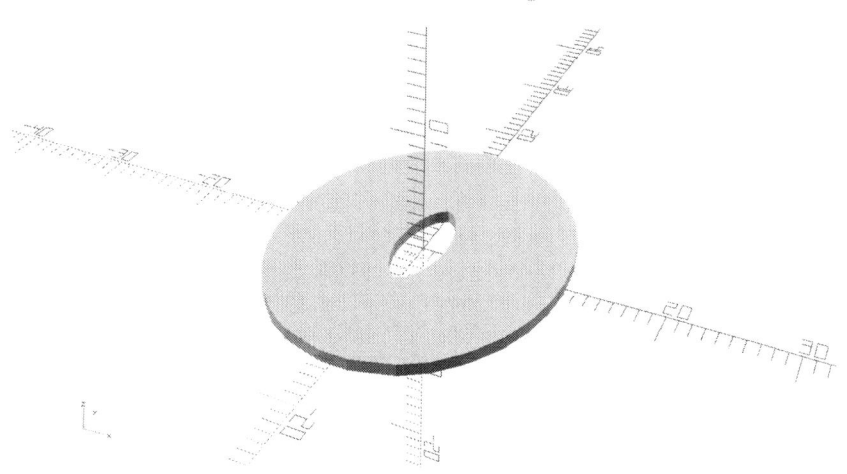

The projection command has one optional parameter named cut, and setting it to true changes the projection to slice through the doughnut where the x,y plane intercepts its volume.

So, instead of projecting all of the doughnut's points to the x,y plane, the cut is more like a lunch meat slicer, where you get your slice for lunch from exactly where the x,y plane passes through the object. I kid you not and there's no baloney here:

```
projection(cut=true)
rotate([0,45,0])
rotate_extrude()
translate([10,0,0])
circle(5);
```

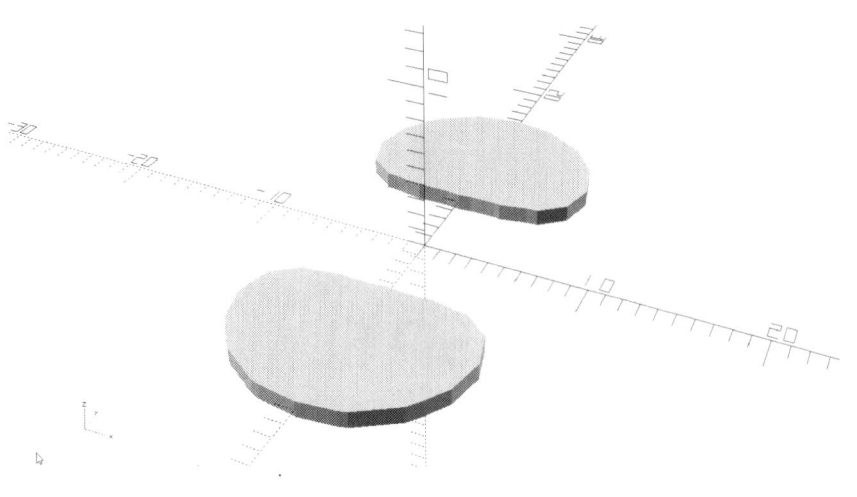

You don't have to settle for slicing your doughnut (or any other object) through its center on the x,y plane. Simply translate an object up or down to position it for the x,y, plane slice to give you what you want.

For example let's shift the doughnut up five units using translate([0,0,5]) and then cut through it. I've colored the result as "salmon" just to stick with our food motif. You might prefer "PeachPuff" (yes, that's a real color) but the thought of a salmon doughnut will help you better retain what you learn about projection(), guaranteed. Sorry about that.

```
color("salmon")
//projection(cut=true)
translate([0,0,5])
rotate([0,45,0])
rotate_extrude()
translate([10,0,0])
circle(5);
```

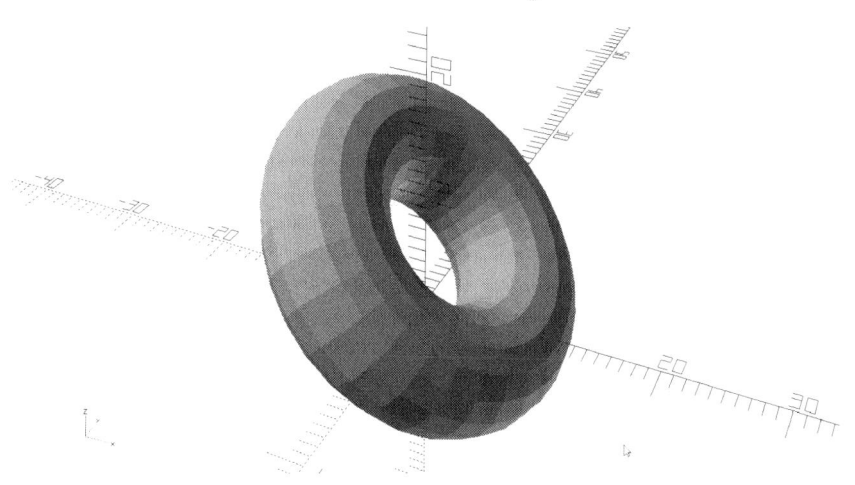

This doughnut has been lifted up by 5 units, ready for the slicer to shave off a slice closer to the bottom of the doughnut. Un-comment the projection() command to complete the process:

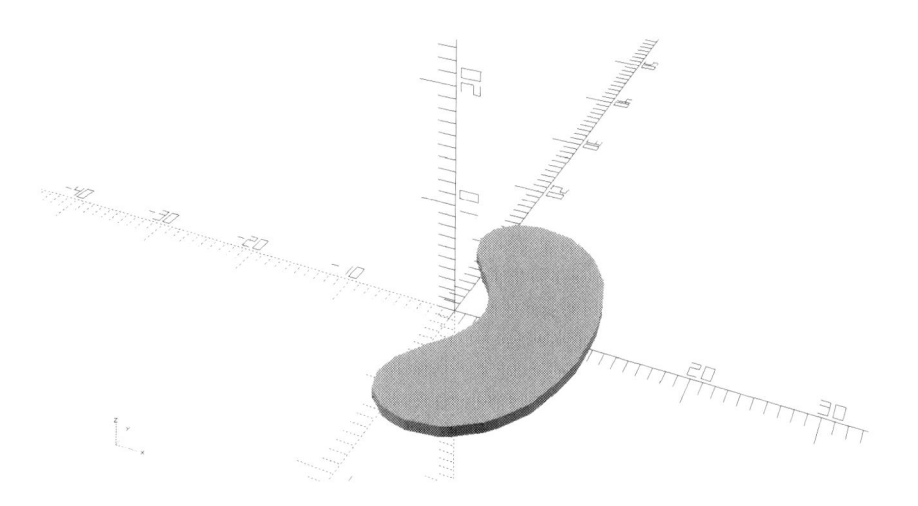

Yummy! Lunch is served.

Seriously though, you can rotate an engineered 3D object around each axis, create a projection 2D image, and later build a document to illustrated your creation from all three sides.

Notes

Notes

Recipe 14

Create the "Holey" Grail

When two or more 3D objects overlap in space, OpenSCAD combines them into one object automatically, using the union() operator by default. For example, both of these two blocks of code create the same "double-bubble" shape by combining two spheres in space into a single object:

```
// Default union()
translate([7,7,0])
sphere(12);
translate([15,9,9])
sphere(7);


union() {
  translate([7,7,0])
  sphere(12);
  translate([15,9,9])
  sphere(7);
}
```

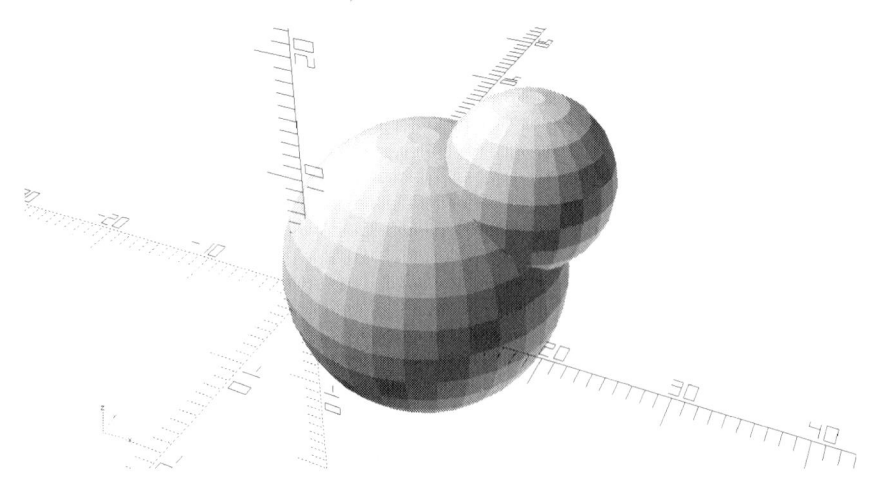

Feel free to use the union() operator if it makes your code more clear or maintainable.

OpenSCAD provides two other boolean operators in addition to union(). The difference() operator creates the first shape in its block of code, and then erases from space everything else that follows. A quick change to the code above causes the second sphere to be deleted from space, especially apparent for those parts where it overlaps in space with the first sphere:

```
difference() {
   translate([7,7,0])
   sphere(12);
   translate([15,9,9])
   sphere(7);
}
```

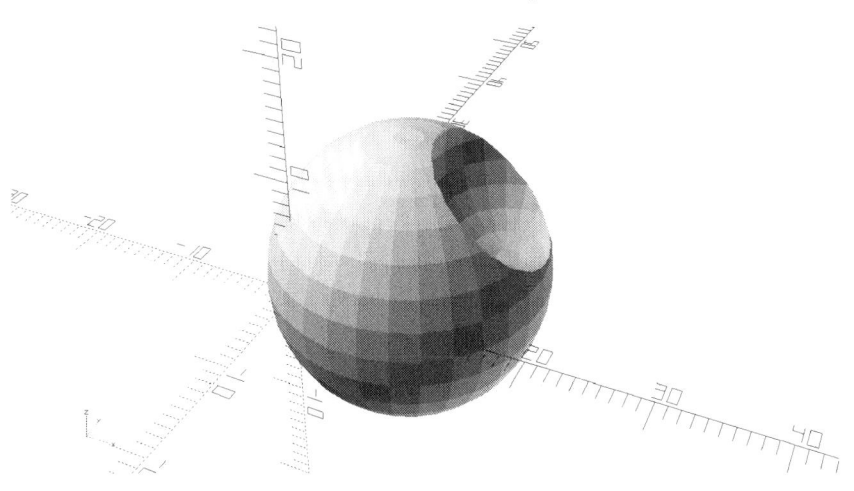

The difference() operator acts like some hungry person is taking bites out of your meatball. Or, perhaps on a grander scale, we've taken the first steps to creating a Death Star.

The third operator is intersection(), and this one erases everything in space created in its code block except where the shapes do overlap or intersect. Another slight modification to the code above results in leaving a hunk of shape where the two spheres cross paths in space:

```
intersection() {
  translate([7,7,0])
  sphere(12);
  translate([15,9,9])
  sphere(7);
}
```

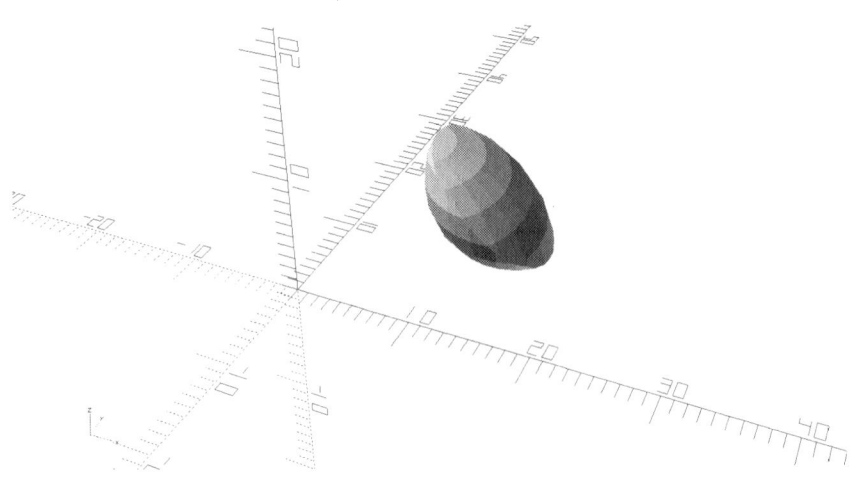

Picture this as the piece of meatball someone earlier bit out of the bigger meatball, where most of the two original meatballs have been slipped under the table to the dogs.

Use your mouse to spin the results around to see how the two curved faces have curvature that matches the surfaces of the two original meatballs, I mean spheres.

Using Cylinders for holes

I used spheres for the examples above, because an earlier recipe already introduced this shape. A much more common use of the difference() operator is to simulate drilling holes through shapes, for bolt holes and similar constructions. The 3D cylinder shape provides the right tool for this job when combined with difference(). The cylinder shape will be presented in a recipe coming up very soon.

Now about that Holey Grail...

The difference operator first creates one shape, then subtracts all the shapes that follow in its block of code. The following example creates the Holy Grail from an earlier recipe (I've "black boxed" its code into a module), then punches some holes in it using spheres. It's all in fun of course, but a little silliness helps you effectively visualize and remember how the difference() operator works!

```
module holy_grail() {
  thick=1.5;
  points=[
  [0,thick],
  [10,0],
  [10,thick],
  [thick,thick+thick],
  [thick,20],
  [12,30],
  [15,45],
  [15-thick,45],
  [12-thick,31],
  [0,21]
  ];

  rotate_extrude()
  polygon(points);
}


// Holey Grail
difference() {

  // Start with the holy grail
  holy_grail();

  // Make first hole
  translate([7,-7,27])
  sphere(5);

  // Make second hole
  translate([-3,-10,35])
  sphere(7);
}
```

John Clark Craig

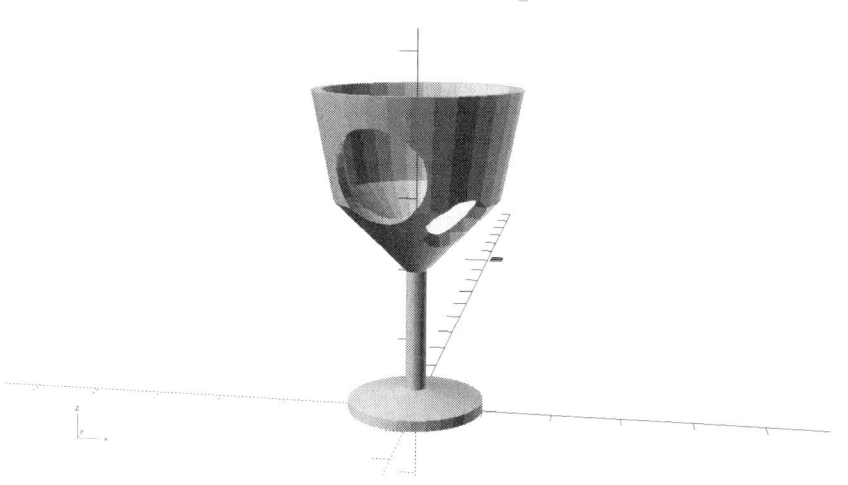

Notes

Notes

Recipe 15

Birthday Candles and Other Common Cylinders

So far we've mostly been creating 3D objects by extruding 2D objects such as circle, square, and polygon. The sphere was introduced in earlier recipes to instantly create a 3D object, and now it's time to explore the cube, cylinder, and polyhedron commands to create other instant 3D shapes. This recipe will focus on the cylinder, a common shape for building up more complex constructions.

The cylinder command provides parameters to control the radius or diameter at each end of the object, its height or total length from end to end, and a center parameter to shift the cylinder such that its center point is on the origin.

As a first experiment, create a birthday candle-shaped cylinder using a minimum of the cylinder's parameters:

```
cylinder(h=200,r=10);
```

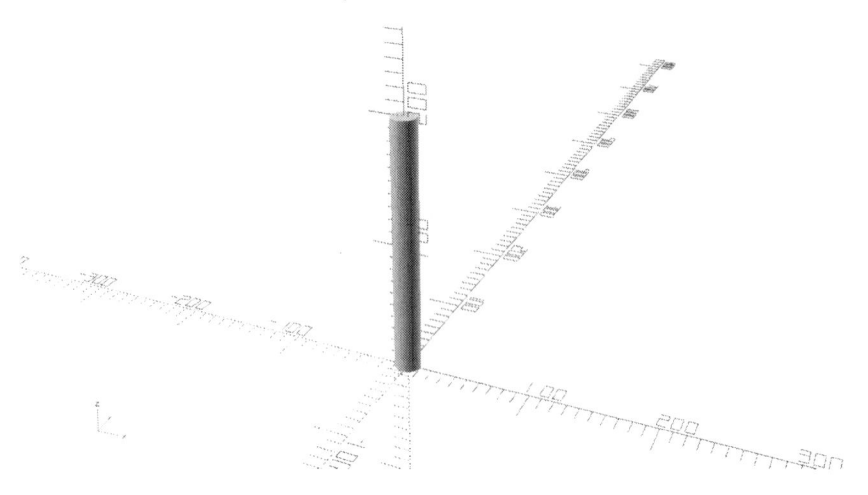

You can leave off the "h=" and "r=" names for these parameters, but I suggest you use them for the cylinder command. There are several combinations of these parameters possible, the order is important if you don't name them, and the added names make the code easier to read and understand in the future. You'll avoid confusion if you name the cylinder's parameters.

The cylinder is created on the z-axis, with its base on the x,y plane, as shown in our first example. Add rotation and translation commands to place the cylinder where you want, such as on top of the birthday cake created in a previous recipe.

Instead of a single "r=" parameter, you can taper the cylinder as desired using "r1=" and "r2=" to control the radius of the cylinder at each end. Let's make a crude Happy Birthday party hat shape using this technique:

```
cylinder(h=200,r1=150,r2=0);
```

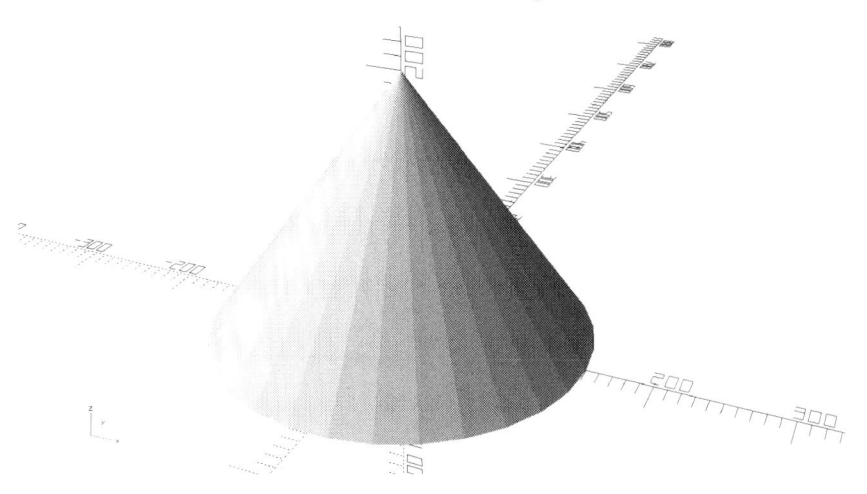

Of course if you spin this thing, you'll see that its flat surface is not going to balance very well on party heads. So, let's use what we learned in the previous recipe for the difference() operator to carve out the inside of this party hat:

```
difference() {
  cylinder(h=200,r1=150,r2=0);
  translate([0,0,-10])
  cylinder(h=200,r1=150,r2=0);
}
```

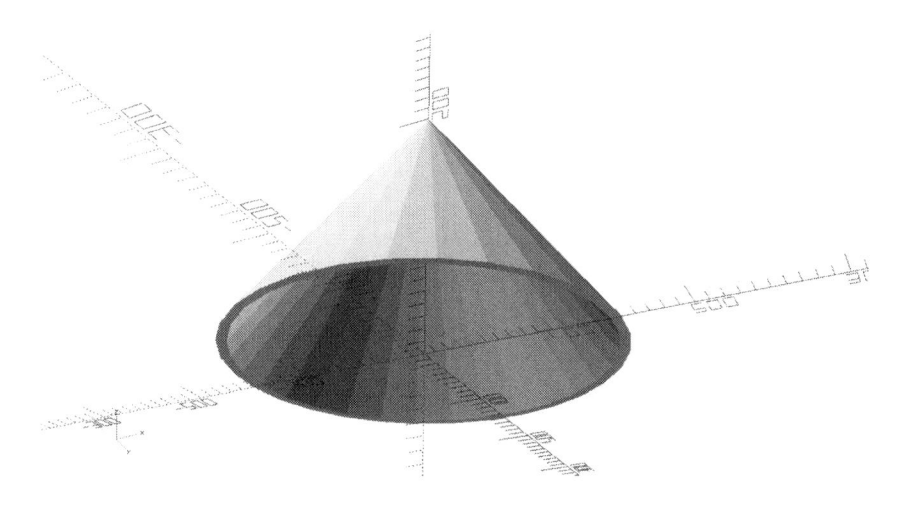

The trick is to use difference() to erase exactly the same 3D shape, except that it is lowered along the z-axis by 10 units, leaving a thin shell. The illustration shows the resulting hat rotated with the mouse so we can take a peek underneath. That should work somewhat better on most birthday party heads.

Pipes and tubes are another common shape in the 3D engineering world. Let's create a hollow drinking straw, perhaps to use with our Holy Grail cup (Warning: it won't work very well with the Holey Grail object):

```
difference() {
    cylinder(h=700,r=11,center=true);
    cylinder(h=705,r=9,center=true);
}
```

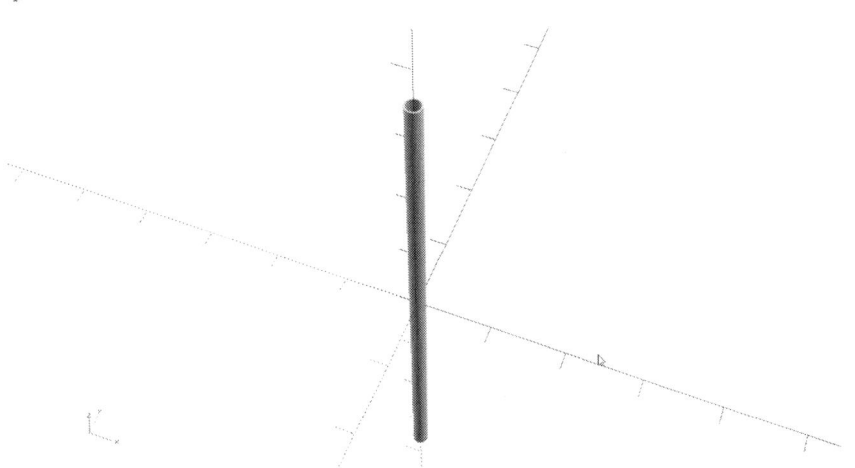

You can carefully set the internal and external radius values to simulate PVC pipes, metal tubes, or party straws.

I've used the "center=true" in this case to shift the straw along the z-axis to center it at the origin. Cylinders are always centered at the origin around the x and y directions.

To make sure the cylinder is hollow, the inner cylinder should be longer, so it "sticks out" of both ends. the difference() operator will then completely hollow out the tube.

Everywhere that a radius value is used in the cylinder command you can optionally use the diameter instead. Substitute "d=", or "d1=" and "d2=" to see how this works. PVC piping, as one example, is designated by internal and external diameter values, so it's more natural to set the diameter values in OpenSCAD in this situation.

Notes

Notes

Recipe 16

Ice Cubes for Party Drinks

An important 3D shape command is cube(), which can be used to instantly create a cube, or any rectangular solid with three edge lengths. In keeping with our party recipes theme, our first example is an ice cube to chill your Holy Grail drink with the straw in it, used to wash down your red Happy Birthday cake with white frosting letters on top:

```
cube(25.4);
```

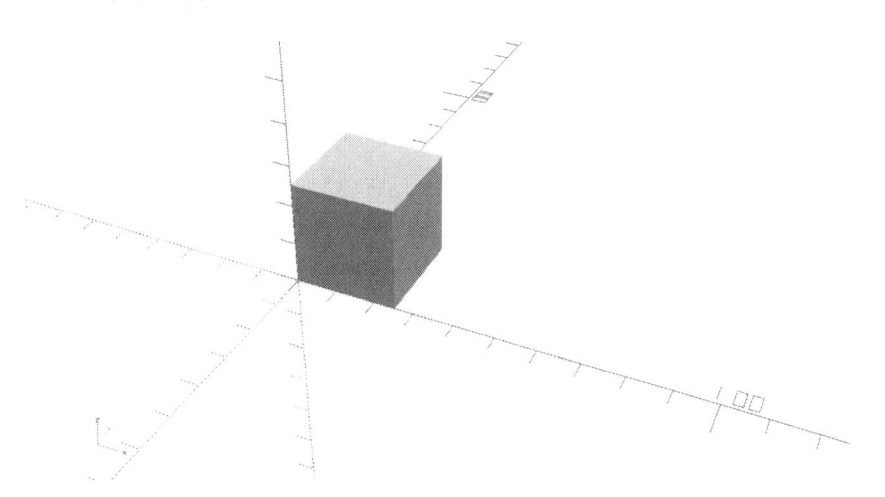

The dimension of 25.4 was chosen to make an important point. There are no units in STL files. If this ice cube were exported to an STL file for printing, it would be sized as a 25.4 units cube, but that could be 25.4 miles if desired (a giant iceberg from Antarctica?) A lot of 3D printing software and hardware assumes mm (millimeters) units, although it really is important to know that STL files are dimensionless. There are 25.4 mm in one inch, so many 3D printers by default would print this as a 1 inch cube. Just pay close attention to your desired units, what your 3D printer expects or assumes, and to how you might set the units for the print job explicitly.

Notice how one corner of the cube is at the origin. Like the 2D square() command, the default is to place a corner at the origin. The "center=true" parameter overrides this action:

```
cube(25.4,center=true);
```

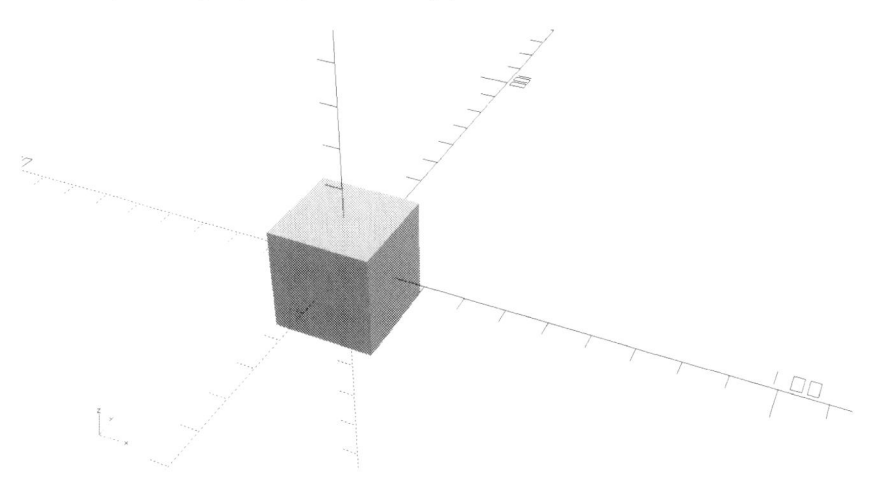

The cube is the same size. It's just been translated along all three axes to move its center to the origin.

To create a rectangular solid of any size, pass it a three-number list instead of a single number for size. Here's a "cube" shaped 30 units along x, 60 units along y, and 90 units tall along the z-axis. Again, this one is centered rather than having one corner at the origin:

```
cube([30,60,90],center=true);
```

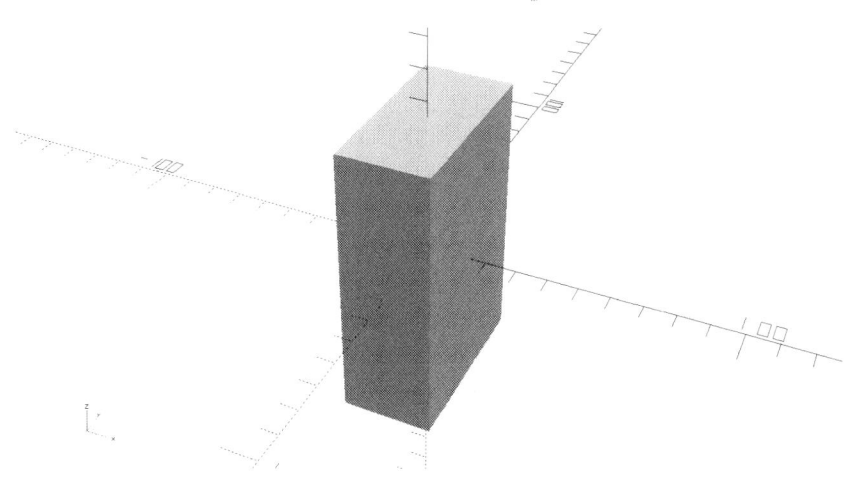

Let's now combine a few concepts from previous recipes to create a hollow box, to store flour in for our doughnuts. We'll create a rectangular solid, then hollow it out using the difference() operator:

```
size = 65;
height = 100;
wall = 3;

difference() {
  cube([size,size,height]);
  translate([wall,wall,wall])
  cube([size-2*wall,size-2*wall,height]);
}
```

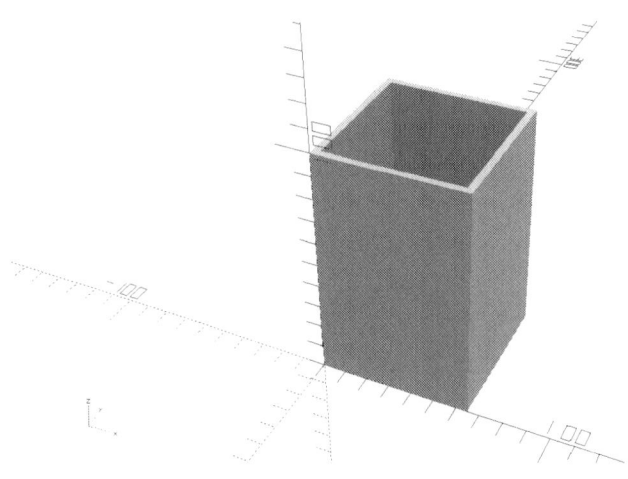

It's always very important to pay close attention to the details. For example, this box has a wall thickness set with the variable named wall, so we shift the second box along each of the three axes by this amount. That moves the interior corner a distance away from the origin by wall units in each direction. The important detail is this interior box needs to be twice the wall thickness smaller than the outer box, in order to get the wall to appear equally on all sides.

Also notice that the order of math operations in OpenSCAD is the same as in almost all programming languages. The math statement "size-2*wall" does the multiplication first, then the subtraction, resulting in the size decreased by twice the wall thickness, as you probably expected.

The interior box's height was created the same as for the exterior box. This extra height (the bottom of the final result is also "wall" thick) guarantees the top of the resulting box will be open.

Now here's a real challenge! Let's create a lid for the box comprised of two rectangular solids, both of "wall" thickness, glued face to face with the default union() operator, and sized such that one square fits just inside the box top, and the other rests on the top edges of the box. Finally, move this lid around in space to position it near the top where it can be spun around and inspected.

Here's one way to accomplish this challenge, but do try experimenting with the code. Make the lid thicker, move it to a new position in space, perhaps just above the box opening, or even put a knobby handle on its top if you feel so challenged!

```
size = 65;
height = 100;
wall = 3;

difference() {
  cube([size,size,height]);
  translate([wall,wall,wall])
  cube([size-2*wall,size-2*wall,height]);
}

translate([50,50,170])
rotate([45,145,0])
union() {
  cube([size,size,wall]);
  translate([wall,wall,0])
  cube([size-2*wall,size-2*wall,2*wall]);
}
```

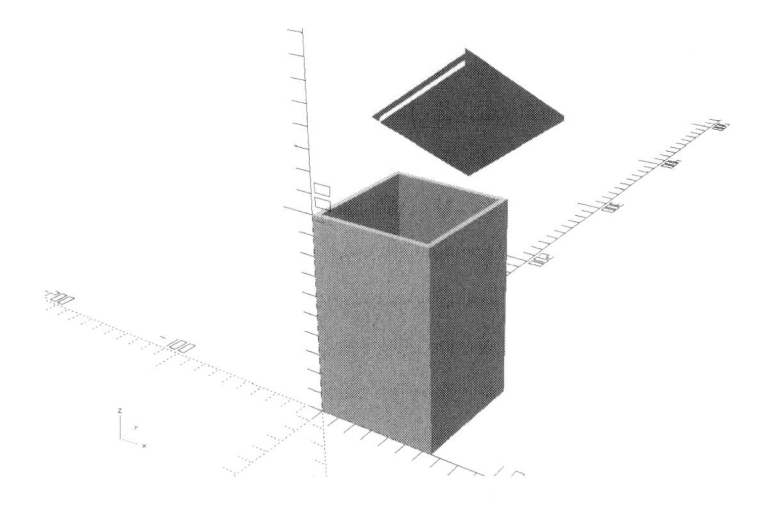

Notes

Notes

Recipe 17

Polyhedron Souffle

The most flexible, but the most complicated, instant 3D shape command in OpenSCAD is polyhedron(). I put the word "souffle" in the title of this recipe so you'd bake it very carefully and not let it collapse. Seriously though, the polyhedron() command demands our full attention to understand, so I'll drop the food references for a bit and focus on a clear explanation. Once we get this thing down maybe we can celebrate with a good old fashioned fun food fight, but not until then!

The polyhedron() command takes two lists and makes a solid. Actually, there's a third parameter named convexity, but it's not used much. I suggest ignoring it until it's needed, then look it up starting with the Cheat Sheet link into "polyhedron()".

The first list is a series of space points. Each space point is itself a list of three numbers representing an x,y,z point in space, so the first list is technically a list of lists. To form a solid in space you need at least four points, so let's create a simple example list of points that will define a volume. Here's a list of four points, three along each of the axes, and one at the origin:

```
points=[[0,0,0],[10,0,0],[0,15,0],[0,0,20]];
```

The second list defines the faces on the solid that these points define. The second list is a list of face points, with some very specific rules you need to follow. Each face contains indexes pointing to the points in the first list, starting with zero. So, the four points in our points list are referenced with indexes 0 through 3.

An example face from the above points might be [0,1,2] which indicates the triangle in space formed by connecting the points [0,0,0], [10,0,0], and [0,15,0]. The solid we're trying to form is comprised of four triangles, and the first one just described lies in the x,y plane.

The second list, in this particular instance, contains a list of four faces, with indexes indicating which points in the first list make up the corners of each face. Here's our second list for the faces:

```
faces= [[0,3,1],[1,3,2],[2,3,0],[0,1,2]];
```

Take some time to make sure you understand how this second list contains numbers pointing into the first list.

The faces may contain more than three points each. For example, the faces on a cube each are defined by four points, always lying in the same plane in space. Of course you can always break up a face into multiple triangle faces if desired, but it's not a requirement as long as all the points used to define a face are in the same flat plane in space.

There's one other very important rule to know and follow. The order that the points in each face are listed is critical. The best way to keep the order correct is to use the right-hand rule as you walk around the points on a face. In your imagination place your right hand such that your curled fingers are walking around the face in the same order as the points are listed. Your thumb should then be pointing towards the inside of the volume. If it's pointing out from the 3D polyhedron shape, reverse the order the points are listed in that face.

Here's the full code listing to create the polyhedron defined by our two lists:

```
points=[[0,0,0],[10,0,0],[0,15,0],[0,0,20]];
faces= [[0,3,1],[1,3,2],[2,3,0],[0,1,2]];
polyhedron(points,faces);
```

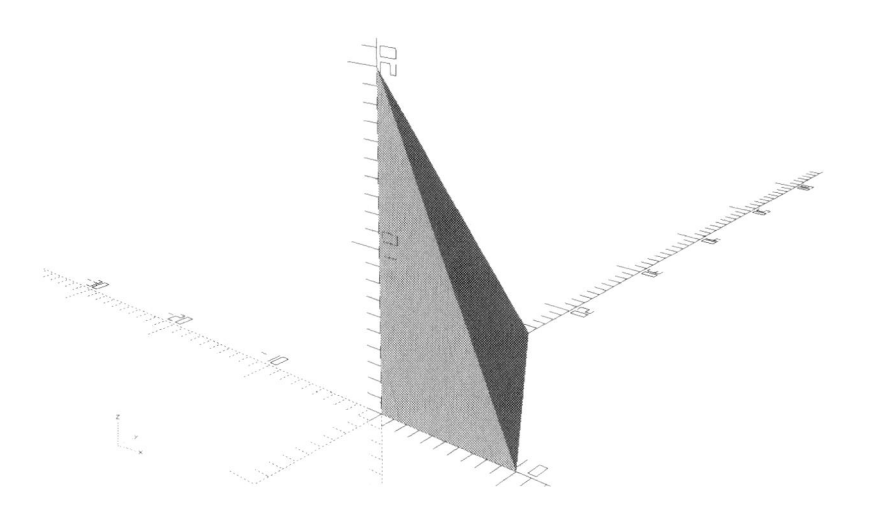

Again, take some time to study the points, and the lists of those points in the faces list, until you understand how it all works to create the shape.

The polyhedron() command allows creating just about any 3D shape you can imagine, as long as it can be broken down into flat faces. It's very powerful, very flexible, and yes, very complicated. There are times when it is indispensable though.

The official online OpenSCAD manual provides more examples of the polyhedron() command, and studying it along with this recipe can help you get a more solid understanding of its rules of construction, and how it can best be used.

And now, here's something entirely different, sneaking up behind you:

```
color("plum")
cylinder(h=30,r1=80,r2=100);
color("mintcream")
linear_extrude(35)
translate([-43,-35,0])
text("\u03c0",size=90);
```

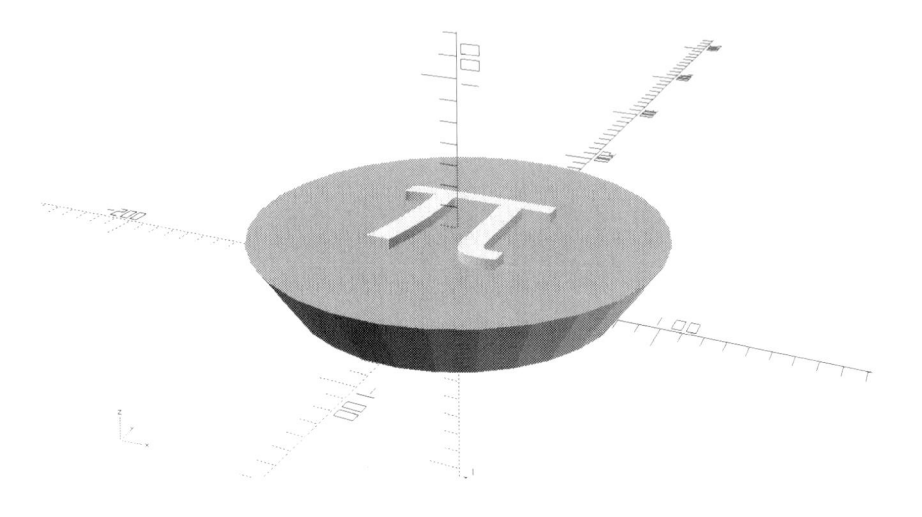

... perfect for starting our food fight!

Notes

Recipe 18

After-Dinner Mints and Toothpicks

There a few more modifier commands that are useful when creating 3D objects using OpenSCAD. This recipe covers two related commands, scale() which stretches or shrinks objects in each of the 3 directions independently, and resize() which does the same thing with absolute dimensions instead of relative.

First we'll create an after dinner mint by creating a sphere that is then stretched into shape:

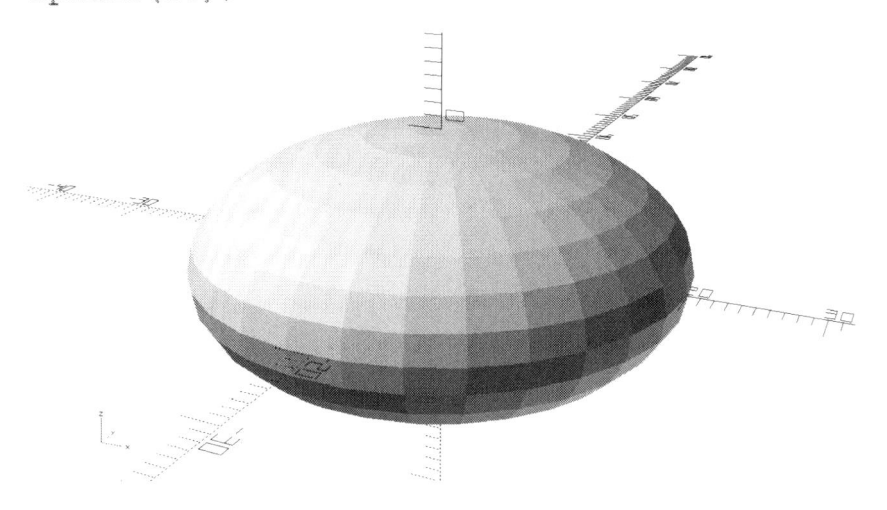

```
scale([2,2,1])
sphere(10);
```

Notice the sphere has been stretched by a factor of 2 along both the x and y axes. The z-axis coordinates are scaled by a factor of 1, which causes no stretching.

Consider this similar code:

```
scale([1,1,.5])
sphere(20);
```

This sphere starts out twice as large as in the first case, but the scaling in all three directions has been cut in half. This scaling shrinks the sphere in the z direction instead of stretching in the x and y. If you try this code you'll see the dinner mint shape and size results are identical to the first example.

The resize() modifier command doesn't multiply each dimension by a factor, it simply resizes the object to the dimensions given. You can stretch a sphere into a mint shape just as easily with this command, but you'll pass it the actual sizes desired rather than multipliers.

Let's go a little extreme and resize a sphere into a long, thin, toothpick. Start with any size sphere, then resize it to exactly 30 units long and only 1 unit in the other directions. Here's the result:

```
resize([30,1,1])
sphere(10);
```

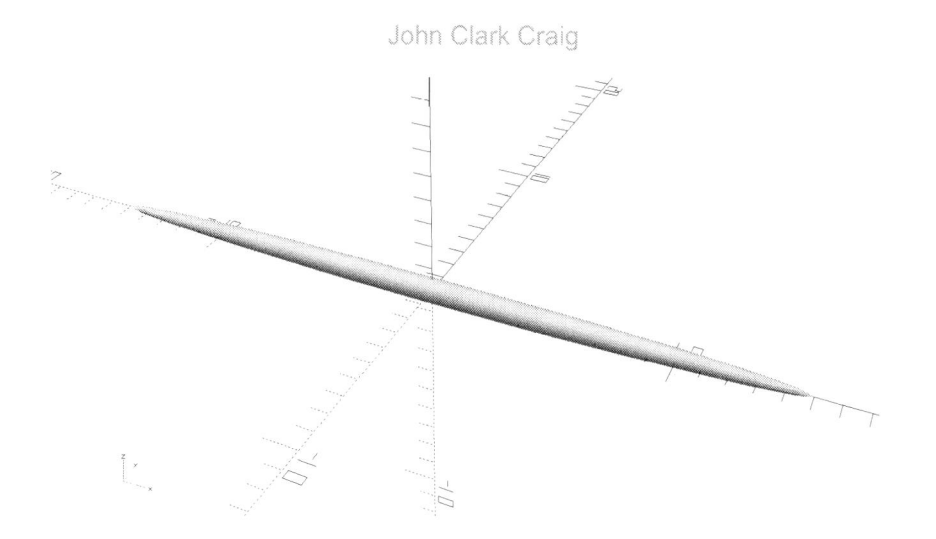

You can stretch other shapes using either the scale() or resize() commands, but it's often easier to simply start with the size desired. However, if you scale() or resize() after rotations or other modifying commands, the results can be dramatic, different, and potentially quite useful.

Here's an example where the same size cube is processed in four different ways using scale() with and without rotation(). Starting at the origin the first cube is created as a reference object. The next cube starts out the same size, but it is scaled along the x-axis direction by a factor of 3, resulting in a rectangular solid three times as long as the first cube. The third cube is rotated about the y-axis by 45 degrees, again just to be compared with the last cube, which is first rotated the same 45 degrees but then also scaled by the same factor of 3. As you can see, the result is a unique double-wedge shape:

```
// Unmodified cube
cube(4);

// Stretched along x by 3
scale([3,1,1])
translate([0,10,0])
cube(4);

// Rotated 45 degrees
translate([0,20,0])
rotate([0,45,0])
cube(4);

// Rotated and stretched
translate([0,30,0])
scale([3,1,1])
rotate([0,45,0])
cube(4);
```

Notes

Recipe 19

Use a Recipe File Box

As a master chef, recipes for common parts of complete meals are best stored away on their own cards in a box somewhere, for easy reference. For example, you might have a card for "Cheese sauce" that you pull out and use when making scalloped potatoes, macoroni, or drizzled over brocolli, or whatever.

This recipe demonstrates how to store away common modules in a separate file, ready to include with ease in more complicated recipes. It's exactly the same as when cooking, only different.

Here's a slight modification of our red "Happy Birthday!" octagon cake, where the main baking code is enclosed in a module. This makes it easy to create multiple cakes, and it makes it easy to store this whole cake thingy into a separate file for easy use in other recipes.

Use File - Save As from the menus to store the following code in a file named "cake.scad":

```
function cartesian(radius,angle)=
  [radius*cos(angle),radius*sin(angle)];

module triangle(p1,p2,p3) {
  polygon([p1,p2,p3]);
}

module regular_polygon(sides=3,radius=1) {
  for(n=[0:sides-1]) {
    p1 = cartesian(radius,n*360/sides);
    p2 = cartesian(radius,(n+1)*360/sides);
    triangle([0,0],p1,p2);
  }
}

module cake() {

  // Parameterized values
  poly_edges = 8;
  poly_radius = 20;
  poly_height = 10;
  text_scale = 0.3;
  text_translate = [-17,-1.2,poly_height-.1];
  text_thick = .4;
  font="Liberation Sans";

  // The red "stop sign" cake
  color("red")
  linear_extrude(poly_height)
  regular_polygon(poly_edges,poly_radius);

  // The verbal frosting on top
  color("white")
  translate(text_translate)
  linear_extrude(text_thick)
  scale([text_scale,text_scale,1])
  text("Happy Birthday! \u263A",font=font);
}

//cake();
```

Be sure to include the last line containing the commented out command "//cake();". This will be explained very soon.

Now start fresh and make a new project named "two_cake.scad" and add these lines to it:

```
include <cake.scad>;

translate([-22,0,0])
cake();

translate([22,0,0])
cake();
```

This code first loads all of the cake code when it encounters the include command, and then processes the lines as though they were in the current program space. All of the code in that file is now in modules and functions, so nothing actually happens until the cake() commands are encountered. Two cakes are then created side by side along the x-axis.

Now go back and remove the comments from the last line of the cake.scad file. Run the code above to include the updated cake code and this is what happens:

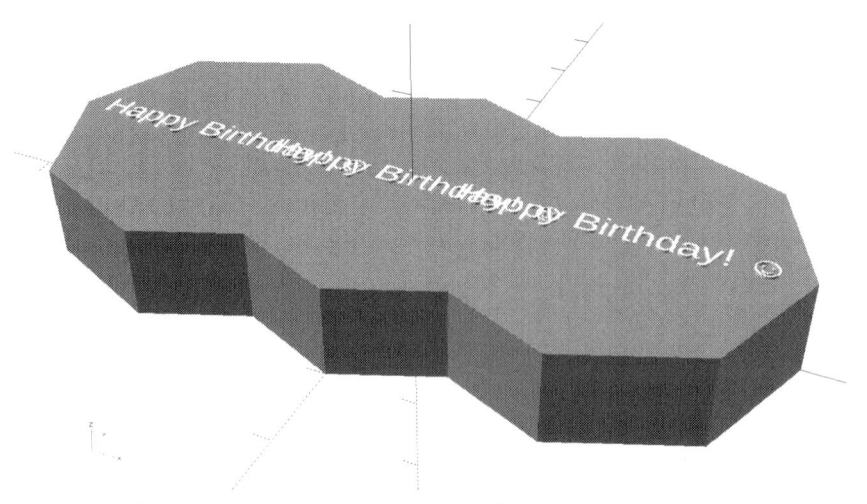

The cake(); command in the included file goes ahead and creates the original version of the cake, located at the origin. This is kind of a mess, but if unexpected extra guests show up, why not.

There are two ways to fix this situation. Either comment out the cake(); command in the included file, or if you like leaving debugging code in your module files, such as the cake(); command, then replace "include" with "use" in your main code. The "use" command loads the file but does not process any commands in the file. All modules and functions are loaded and ready to use, but no code in the loaded file is actually run as the file is loaded.

To see how this works, leave the cake(); command uncommented, and use this code in your main program:

```
use <cake.scad>;

translate([-22,0,0])
cake();

translate([22,0,0])
cake();
```

The result will be the two cakes, and the original cake() command in the "included" file will be ignored. It really is a matter of taste as to which development mode your prefer - neat and clean library files with only functions and modules, or testing code built in that makes standalone debugging and testing easy.

Splitting common code out into library files of modules and functions as shown in this recipe provides a very good way to handle the construction of complex projects comprised of many shapes and parts. And it makes it easy to mass produce your cakes as your party size gets out of hand.

The next recipe explains the mirror() operator, and having our cake in a separate file will shorten and simplify the explanation considerably.

Notes

Notes

Recipe 20

Mirror Mirror on the Plane

I started to name this recipe "Mirrors on a Plane', but that sounded a little too scary. The mirror() operator is nothing to fear.

When I first encountered the mirror() command, it threw me for a loop, and it took me some time to get a handle on how it works. It really is easy, but understanding clearly the concept of how to define a plane is the key.

The mirror() command is passed a three number list that defines a plane that always passes through the origin. The three numbers define a vector, or let's call it a direction arrow, that starts at the origin and points away into space towards the x,y,z point defined by the three numbers.

Now here's the key part. The vector is normal to the plane. This means it is perpendicular to its face. Given a vector starting at the origin and pointing somewhere in space, there's only one plane that can pass through the origin and be perpendicular to that vector. This is why just three numbers can completely define a plane for the mirror() command.

When the mirror() command is applied to an object (it modifies the object just like translate and rotate), all points on that object are moved to the other side of the plane. The result is a "mirror image" of the object as far away from the plane as the original object, but on its other side.

An example will help. Imagine a plane defined by the y and z axes. Any point on this plane will have an x value of zero. Think of this plane as a mirror standing on edge facing to the right, and passing through the origin. One vector that completely defines this plane is [1,0,0], because that vector points away from its face. Another vector for the same plane could be [100,0,0] because it's the direction that counts, not the magnitude.

Let's create our cake shifted away from the origin along the x-axis, and then let's create a second identical copy only mirrored using the y,z plane as described:

```
include <cake.scad>;

translate([22,0,0])
cake();

mirror([10,0,0])
translate([22,0,0])
cake();
```

As you can see, the text is reversed where the second copy of the cake has been mirrored into the space on the other side of the y,z plane. Everything is backwards, just like looking in a mirror.

As a second example, let's build an ice cream scoop (or perhaps it looks more like a melon baller?) and then have some fun with mirrors:

```
module scoop() {

    rotate([0,90,0])
    cylinder(h=30,r=5);

    translate([38,0,0])
    difference() {
        sphere(10);
        translate([0,0,4])
        sphere(9);
    }
}

scoop();
```

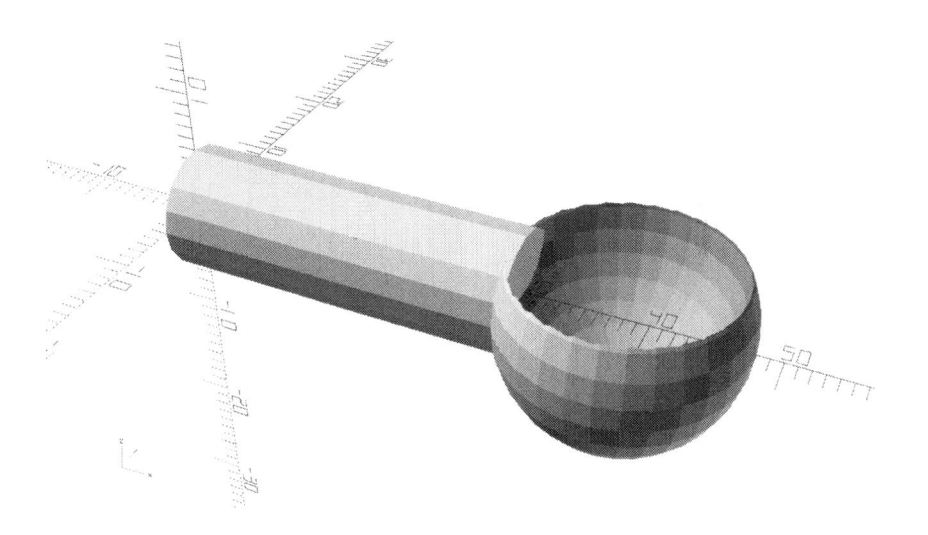

As programmed, the scoop lies along the x-axis. Here's one way to modify the calling code to create three copies of these scoops in a solid pinwheel layout, kind of a weird start towards a kitchen fidget spinner:

```
scoop();

rotate([0,0,120])
scoop();

rotate([0,0,240])
scoop();
```

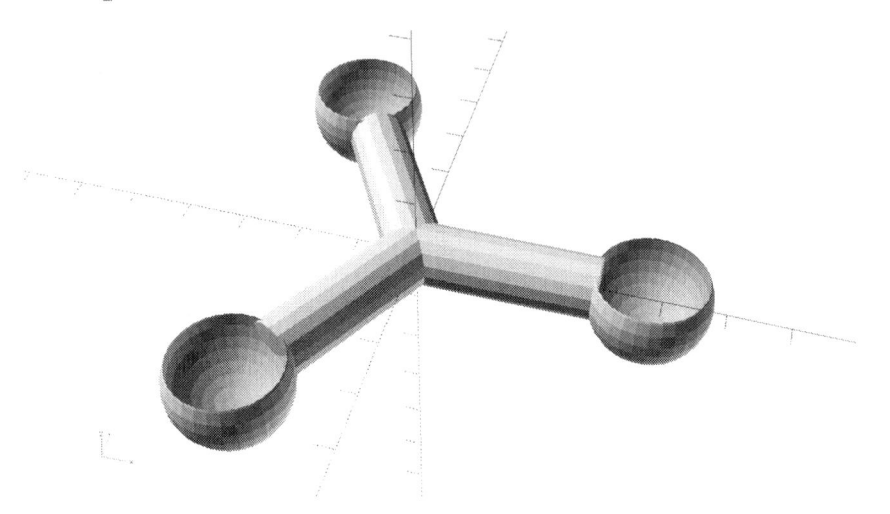

We've already used the rotate command in several recipes. We can accomplish exactly the same goal using mirror() commands instead:

```
scoop();

mirror([sqrt(3),-1,0])
scoop();

mirror([sqrt(3),1,0])
scoop();
```

This code creates exactly the same 3-ended scoop object as before.

Each mirror() command defines a plane using its vector. Both planes are vertical, as there's no z component pointing the way, one facing 30 degrees below the x-axis, and the other facing 30 degrees above the positive x-axis. The mirrored positions of the scoops end up exactly as before.

Notice the square root function in the mirror() command vectors above. OpenSCAD has a long list of common mathematical functions useful for 3D analytical geometry, such as creating a vector in space at known angles. Refer to the Cheat Sheet for a list of the others. Like everywhere else in OpenSCAD, the trigonometric functions all assume angles are expressed in degrees.

Both examples so far have used vertical mirror planes to mirror our scoop around in basically just the x,y plane. Here's an example where we use mirror to angle the first scoop down away from the x,y plane. The second scoop is created the same as the first, but a second mirror() command reflects it to a position behind the y,z plane.

```
mirror([1,0,1.3])
scoop();

mirror([1,0,0])
mirror([1,0,1.3])
scoop();
```

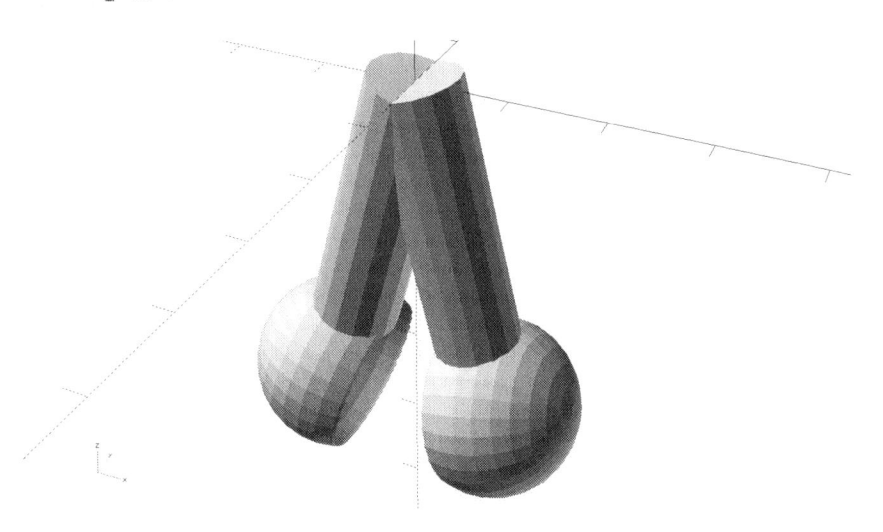

You could hang these from the rear bumper of your chef-mobile, to show the world you are the super-duper-double-scooper in the kitchen, but I would strongly advise against this idea. Just saying.

Notes

Recipe 21

Popcorn and Other Hulls

An interesting and powerful OpenSCAD command is hull(). Hull operates on 3D objects to kind of "shrink wrap" them. That is by far the easiest way to describe the action, and a few examples will make it all clear. There's actually no popcorn involved, but the hull() operator has about the same entertainment factor.

First, create two cubes side by side:

```
cube(10);

translate([15,0,0])
cube(10);
```

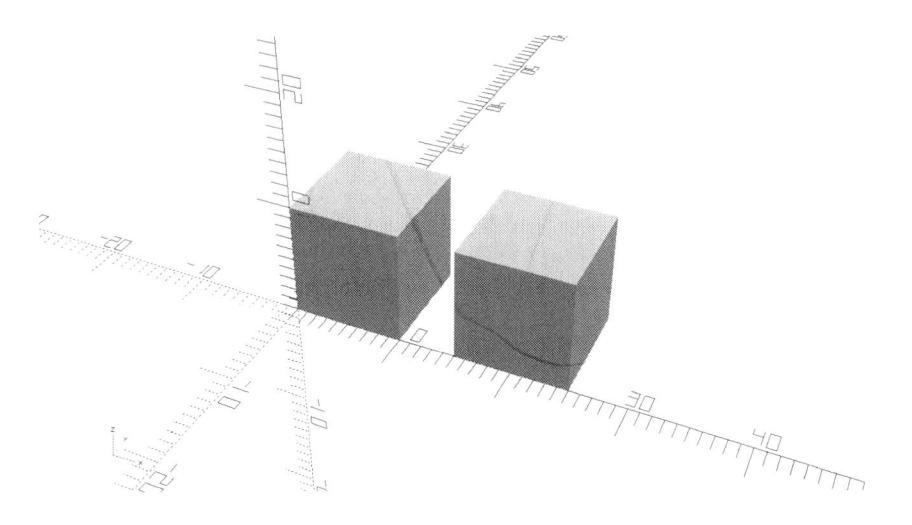

If we wrap both of these cubes in a hull() operator, a convex hull is shaped around them (that's the graphics science terminology), resulting in a rectangular solid containing the original cubes at each end:

```
hull() {
  cube(10);
  translate([15,0,0])
  cube(10);
}
```

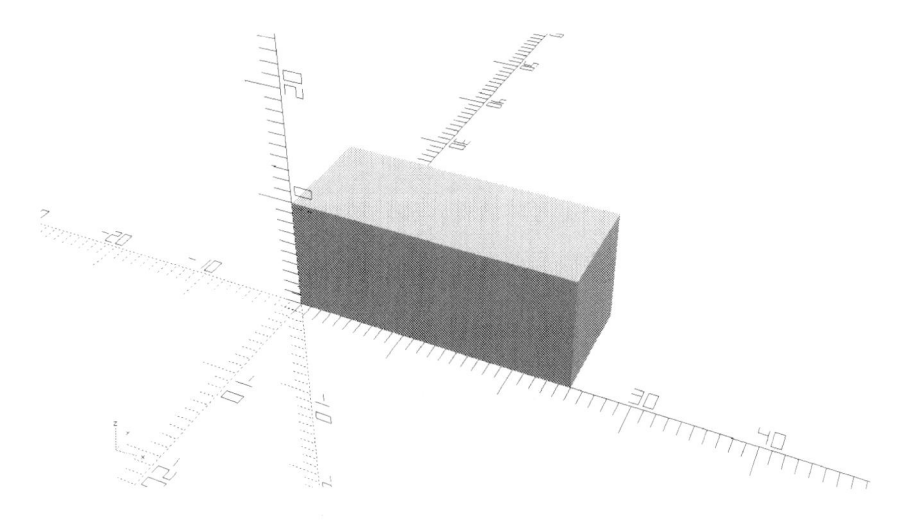

If we translate the second cube to a new location on the x,y plane, the action becomes a little more interesting:

```
hull() {
  cube(10);
  translate([15,15,0])
  cube(10);
}
```

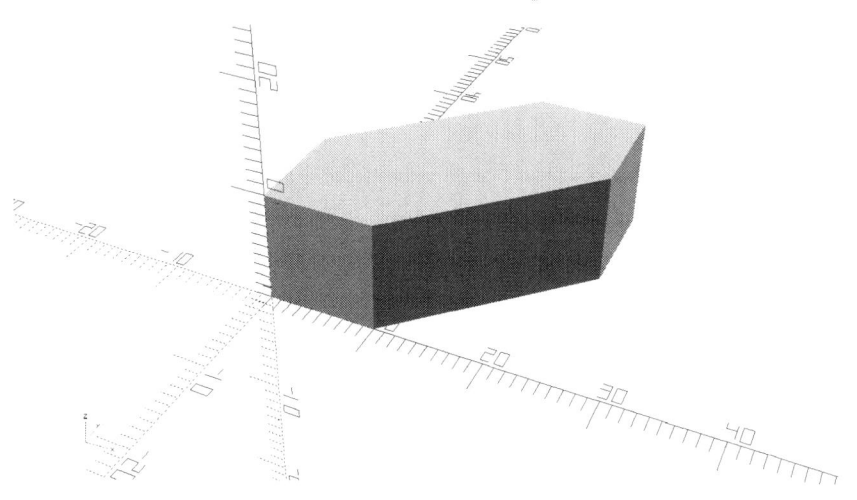

Translating the second cube up in the z direction adds yet another crystal-like structure to the hull-wrapped shape. You can still see several faces of the original cubes, but the faces are connected with surfaces with the minimum area possible:

```
hull() {
  cube(10);
  translate([15,15,15])
  cube(10);
}
```

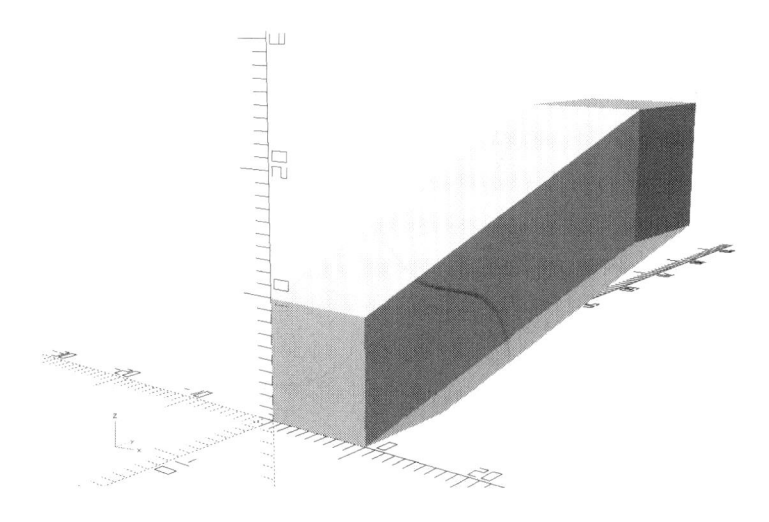

It's fun, and informative, to experiment with the hull() operator. To get back to our food related fun, let's create an ice cream cone. First create a sphere for the ice cream ball, and another smaller sphere below the first one to define the bottom point of the cone:

```
translate([0,0,40])
sphere(20);
sphere(3);
```

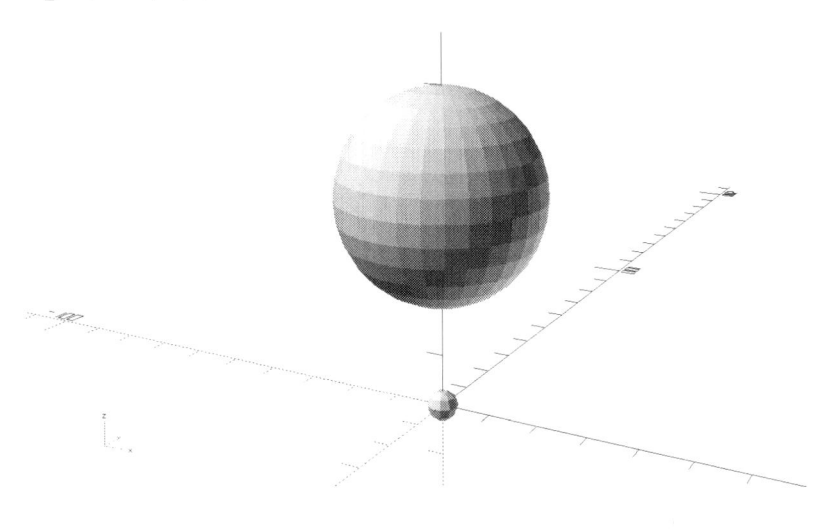

Wrap these two spheres with a hull() and enjoy before it melts!

```
hull() {
   translate([0,0,40])
   sphere(20);
   sphere(3);
}
```

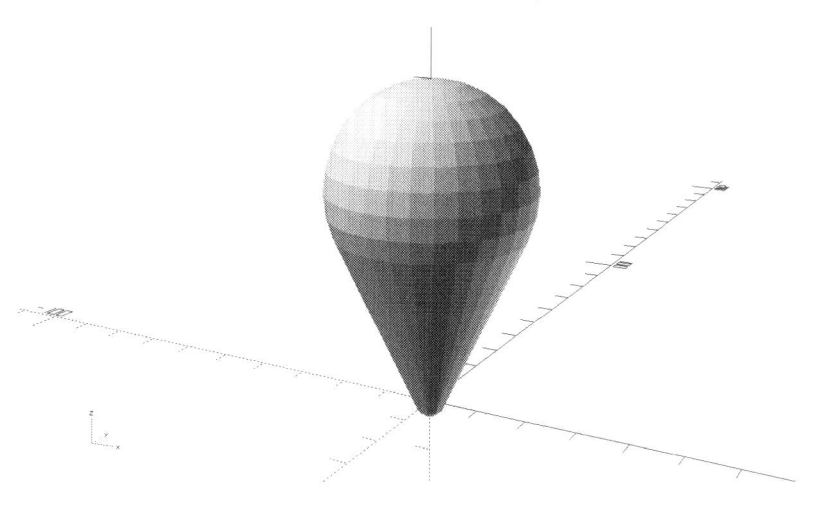

I've seen better ice cream servings (this actually looks more like a hot air balloon), but it's the cone-cept here that's important.

The hull() command wraps anything and everything into the tightest, shrink-wrapped package possible. This can be used in a creative way to round the corners on cubes and boxes. For example, here's a module named rounded_box that creates a rectangular solid with the edges all "sanded down" into round shapes. We'll call the module to create a 50 by 40 by 30 units box with the corners and edges all rounded with a radius of 5 units:

```
module rounded_box(x,y,z,r) {
  hull() {
    translate([r,r,r])
    sphere(r);
    translate([x-r,r,r])
    sphere(r);
    translate([x-r,y-r,r])
    sphere(r);
    translate([r,y-r,r])
    sphere(r);
    translate([r,r,z-r])
    sphere(r);
    translate([x-r,r,z-r])
    sphere(r);
    translate([x-r,y-r,z-r])
    sphere(r);
    translate([r,y-r,z-r])
    sphere(r);
  }
}

rounded_box(50,40,30,5);
```

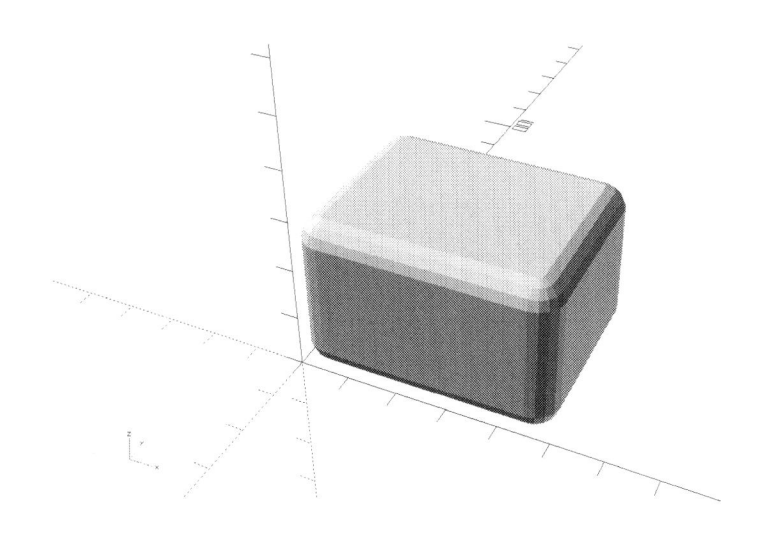

Comment out the hull() command (and its matching brace near the end of the module) to see the eight spheres at each corner of the rectangular solid box:

```
module rounded_box(x,y,z,r) {
  //hull() {
    translate([r,r,r])
    sphere(r);
    translate([x-r,r,r])
    sphere(r);
    translate([x-r,y-r,r])
    sphere(r);
    translate([r,y-r,r])
    sphere(r);
    translate([r,r,z-r])
    sphere(r);
    translate([x-r,r,z-r])
    sphere(r);
    translate([x-r,y-r,z-r])
    sphere(r);
    translate([r,y-r,z-r])
    sphere(r);
  //}
}

rounded_box(50,40,30,5);
```

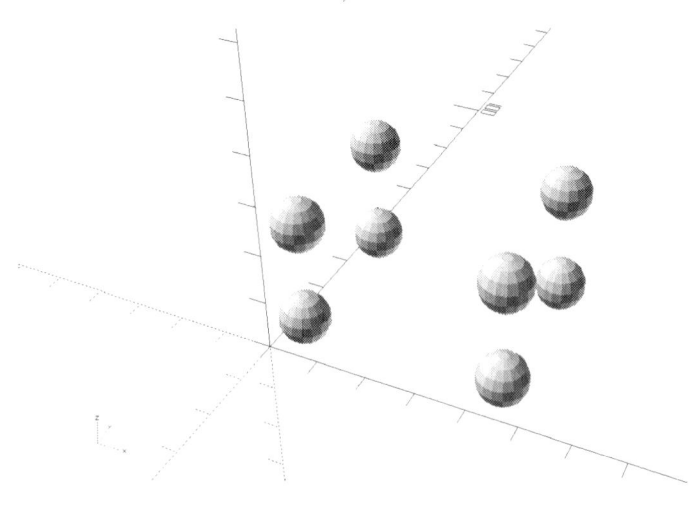

You can round just the edges, leaving the top and bottom surfaces flat and sharp edged, by using the hull() operator on four cylinders instead of eight spheres. Let's call this module round_edges_box(). Here's the same size box, first with the hull() command and its matching brace commented out to display just the four cylinders at the corners of the box:

```
module round_edges_box(x,y,z,r) {
   //hull() {
      translate([r,r,0])
      cylinder(h=z,r=r);
      translate([x-r,r,0])
      cylinder(h=z,r=r);
      translate([x-r,y-r,0])
      cylinder(h=z,r=r);
      translate([r,y-r,0])
      cylinder(h=z,r=r);
   //}
}

round_edges_box(50,40,30,5);
```

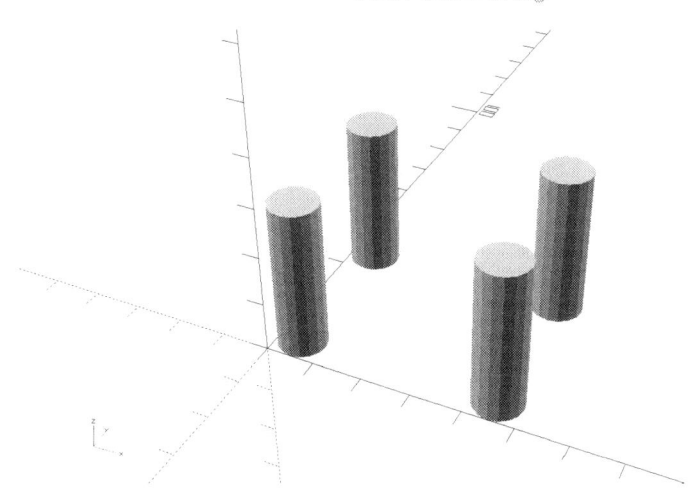

With the hull() enabled we see the box with its edges only rounded:

```
module round_edges_box(x,y,z,r) {
  hull() {
    translate([r,r,0])
    cylinder(h=z,r=r);
    translate([x-r,r,0])
    cylinder(h=z,r=r);
    translate([x-r,y-r,0])
    cylinder(h=z,r=r);
    translate([r,y-r,0])
    cylinder(h=z,r=r);
  }
}

round_edges_box(50,40,30,5);
```

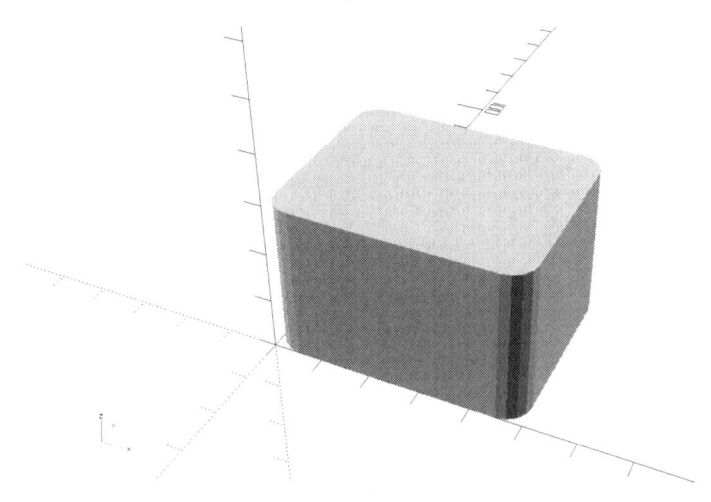

Notes

Recipe 22

Minkowski Mints

In the previous recipe we "shrink wrapped" some shapes using the hull() operator, providing a way to round off the edges and corners of some shapes. OpenSCAD has another operator, named minkowski(), that does something very similar. Let's use it to create a rounded bowl for serving mints.

Be forewarned that both the minkowski(), and hull(), operators can take a long time to complete, especially if $fn is set to a higher number to smooth out the details. I strongly suggest using the default setting for $fn at first, and then slowly increasing its value to make sure your computer can handle the calculations in a reasonable amount of time. When you're ready to smooth out your final model in preparation for creating a printable STL file, then you can afford to wait for the result.

As a first step, create a shallow box and round its edges off by applying a sphere to it using minkowski(). This adds the radius of the sphere to all surfaces and edges. Look close and you'll see the faces have moved past the origin lines by 8 units:

```
minkowski() {
   cube([100,100,30]);
   sphere(8);
}
```

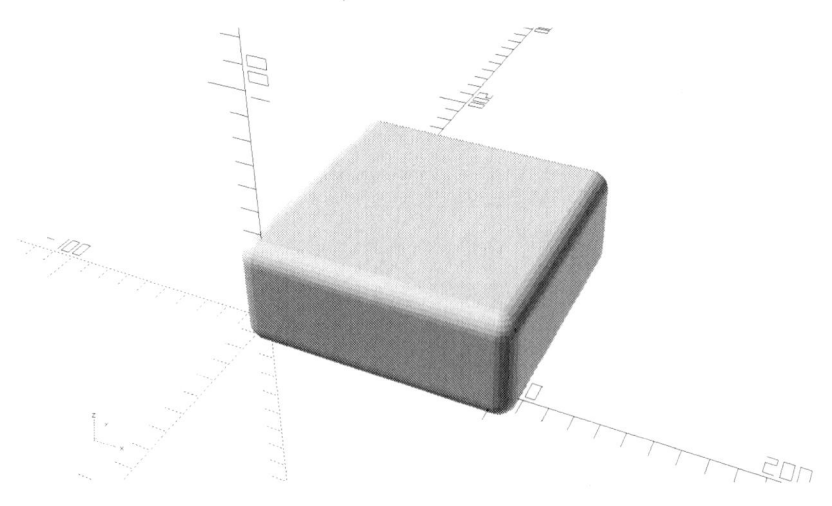

The difference operator lets us carve out the inside of this rounded box. Let's create a smaller version of the same box and subtract it from the original. Notice that the height of this inner box is set very tall, because we want the top of the box to be open. Here's the inner box before applying the difference:

```
translate([10,10,10])
minkowski() {
  cube([80,80,80]);
  sphere(8);
}
```

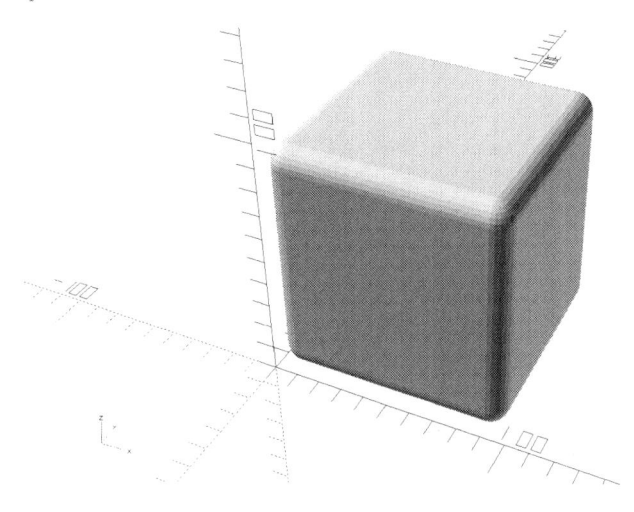

Now we subtract this narrower but taller box from the first to create our serving bowl for mints:

```
difference() {

  minkowski() {
     cube([100,100,30]);
     sphere(8);
  }

  translate([10,10,10])
  minkowski() {
     cube([80,80,80]);
     sphere(8);
  }
}
```

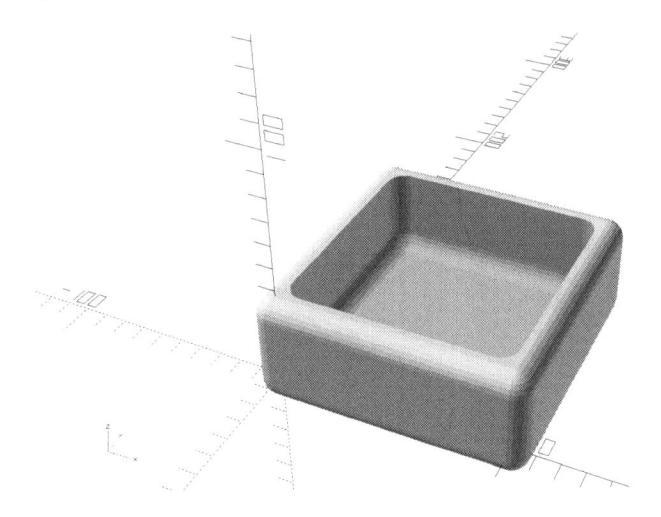

The only non-smoothed edges in the result is along the inner top of the mint box, where the vertical straight edges of the inner box were subtracted from the outer box. The inside edges of this box are nice and rounded.

It's tempting to apply the difference of these two boxes first, before rounding each off using minkowski(). Here's the serving box with no rounding of its edges at all:

```
difference() {

    cube([100,100,30]);

    translate([10,10,10])
    cube([80,80,80]);
}
```

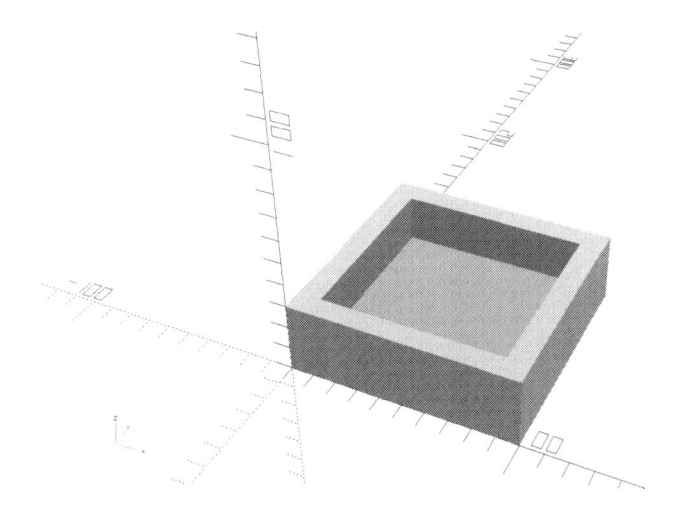

Applying the minkowski() operator with a sphere sized 5 units causes all external edges and corners to be rounded, but leaves the interior edges straight:

```
minkowski() {

  difference() {

    cube([100,100,30]);

    translate([10,10,10])
    cube([80,80,80]);
  }
  sphere(5);
}
```

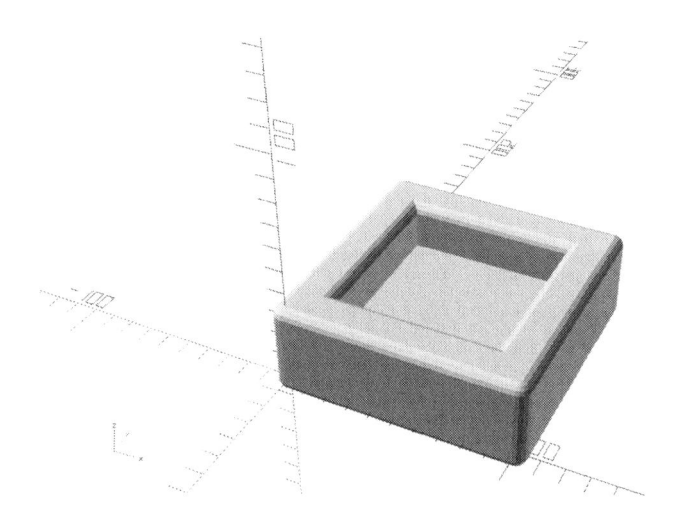

Finally, note that you can apply a cylinder instead of a sphere to cause the minkowski() operator to round just the edges of a shape. Let's try it on our mint bowl for yet another version of the final product:

```
minkowski() {

  difference() {

    cube([100,100,30]);

    translate([10,10,10])
    cube([80,80,80]);
  }
  cylinder(h=5,r=10);
}
```

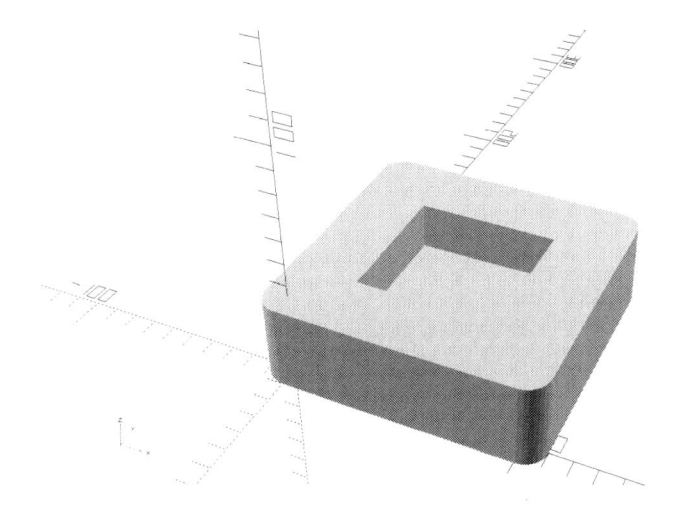

The minkowski() operator adds the radial size of the cylinder in horizontal directions to all faces, and it also adds the height of the cylinder, without rounding, to the top and bottom faces.

The minkowski() operator has enough quirks that I suggest studying the manual pages, and simply trying it out until you get the results you want.

All of the above boxes will work okay, depending on the aesthetics you desire for your mint box, but I suspect the mints will taste about the same no matter what.

Notes

Appendix A

Using OpenSCAD

The online OpenSCAD User's Manual covers all the details of the user interface, and I suggest referring to it when in doubt, or to figure out how to do something out of the ordinary.

https://en.wikibooks.org/wiki/OpenSCAD_User_Manual/The_OpenSCAD_User_Interface

...or use this, for easier typing: https://goo.gl/xgJUqY

My goal in this appendix is to cover the very basics, the stuff I use 99% of the time while interacting with the OpenSCAD environment.

Menus

The File menu is self explanatory if you use your computer much at all. This is where you can create, load, and save files, or exit the program. The Export option outputs the results of your work in several types of file formats, perhaps the most important being STL for sending to a 3D printer.

The Edit menu works with the editor window, which is along the left side of the display in most cases. There are two commands that aren't so obvious, yet they can be handy at times. "Paste viewport translation", and "Paste viewport rotation" lets you paste a three-number list into the edit window at the cursor location. If you insert these lists into the translate() and rotate() commands respectively, the current amount of mouse-caused translation or rotation is available for use.

The Design menu controls how and when OpenSCAD processes your instructions as typed into the edit window. I prefer to check the "Automatic Reload and Preview" option, so the model updates automatically every time I click the icon (or press Ctrl-s) to save the file. Otherwise, the "Preview" option updates the output on a manual basis. The "Render" option uses a different 3D modeling engie to fully render and prepare your instructions before outputting an STL file, whereas "Preview" is somewhat faster but sometimes not as accurate. I almost always use "Preview" while developing, and "Render" for the final product. Be aware that to create an STL file you must use render first.

The View menu is important, and it is definitely worth the time to get familiar with its options. Most of the time, I check only the options for Preview, Show Axes, Show Scale Markers, and Perspective. You might prefer other combinations of the various settings, so try them out to get familiar with the results.

The Help menu accesses documentation on the web, and shows the path to libraries, and a list of available fonts, both of which can be handy at times. The "Cheat Sheet" link is one I use all the time, as the organized list of commands and keywords on that page links directly to the appropriate locations in the main user manual. This saves time and effort when trying to recall syntax details.

Icons and Buttons

There are several icons, or buttons with small images on them, at the top edge of the edit window. These help with editing details, mostly repeating commands in the edit menu. The preview, render, and STL buttons are important for processing your program, although I tend to keep the "Automatic Reload and Preview" option in the design menu set and simply click on the little floppy disk icon to save the file and trigger a preview. Only when I'm ready to fully render and convert to an STL file do I click the other two buttons.

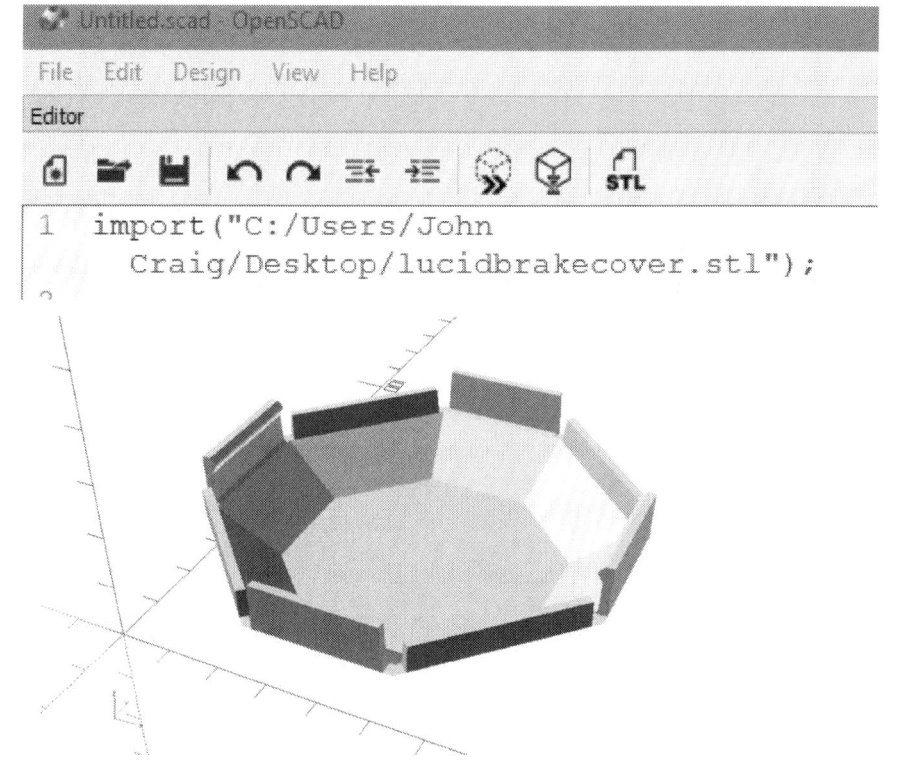

Below the model viewing window is another row of icons. Hover over each to get a text popup description. The first two are repeats of the preview and render buttons. I use two of these all the time, and the rest just occasionally. After zooming and scrolling, or after the model is generated somewhat off screen, I first click on "Reset View", and then immediately on "View All". This causes my objects to be centered in the view, and zoomed appropriately to be completely visible and at a reasonable size. Pan and zoom from there to tweak what you see.

The rest of the icon buttons in this row are repeats of what is available in the View menu. These allow standard rotations for viewing your model from the top, front, and so on. The perspective and orthogonal buttons switch between views where parallel lines in space meet in the distance (perspective), and where they stay parallel (orthogonal).

Mouse Use

The online documentation describes all the ways you can use your mouse. Of course your mouse is handy for menus and buttons, but a very important feature is the ability to translate, rotate, and zoom your model in the viewing window. I hold my left mouse button down and drag to rotate the model, the right button to pan or translate left, right, up, and down, and the scroll wheel to zoom in and out. After a little bit of experimentation you'll find these actions come instinctively and automatically.

Creating STL files

OpenSCAD is great for creating the STL files used by most 3D printers. STL files are readable text, although they can get quite complex in the way they describe all the surfaces and points comprising your models. If you want to know the ins and outs of STL syntax, search on the Internet and you'll find it all there. Just be aware that outputting an STL file is what it's all about if you want to solidify your model into real-world 3D plastic!

OpenSCAD can also import an STL file, and this can be very useful in certain circumstances. For example, a while back I created a snap-on polycarbonate cover for my Lucidbrake invention, using one of the very expensive 3D modeling programs. After expending a lot of time and mental anguish with its learning curve, I ended up with an STL file that a mold maker was able to use. All ended well for getting manufacturing going. Recently, I imported this STL file into OpenSCAD, and recreated a new version of the STL file from there. I'm now in a much better position if I need to make any adjustments to this cover.

There's a trick to importing an STL file that is not at all obvious.

Use the **File>Open** menu to locate an STL file, but you'll need to type "*.stl" into the file name field to override the default *.scad files that are normally the only ones shown. At that point the available *.stl files will show and you can select one to load. Interestingly, opening an STL file into the environment does not immediately show the file in the view window. Instead, a command is entered into your text window to load the file at the next preview.

For example, after I copied my lucidbrakecover.stl file to my desktop, here's the command that was inserted into my edit window when I selected this file:

Animation

A very cool feature of the OpenSCAD environment is the ability to create animations of your models.

Recall that each object can be modified with one or more commands such as translate, rotate, color, and so on. You can animate through a range of values by including the special system variable $t, which cycles through values 0 to 1, in calculations driving these modifying commands.

An example would be to move a small cube from the origin to an x-axis value of 20 with these commands:

```
translate([$t * 20,0,0])
cube(5);
```

Nothing happens until you activate the animation, but that's easy to do. From the View menu click Animation to enable it. Three number fields appear at the bottom edge of the view window. Enter 10 in the FPS field (frames per second), and 50 in the Steps field. The Time field is filled in automatically as the animation proceeds, so don't type anything there.

If you've entered the code lines above, and set the two field numbers as described, the animation should start instantly. Experiment by typing other numbers to change the animation "on the fly".

Here's a screen grab of the animation at 34% of the way through the current cycle:

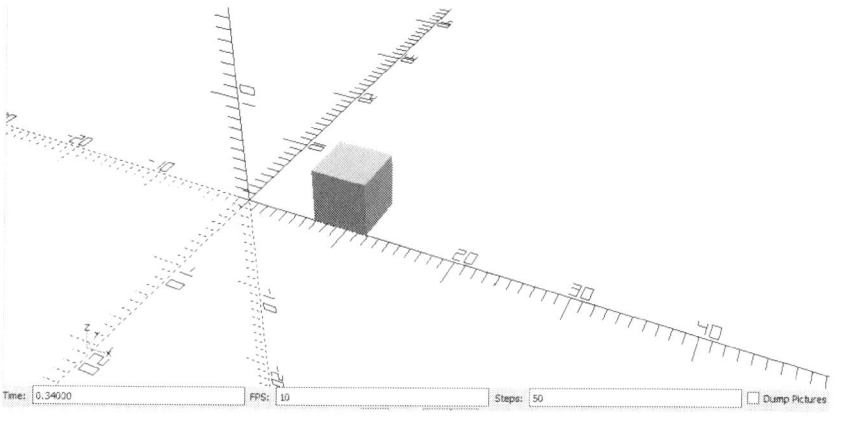

In this example, the cube is recreated 10 times each second, and with each recreation the value of $t is incremented by 1 part in 50, or a decimal value of 0.02. The animation sequence therefore takes 5 seconds total, so the cube slides out to the 20 units point at the end of 5 seconds, and instantly starts over at the origin to start the next cycle.

You can do a lot of fancy animation by creatively using $t in one or more of the modifying commands. As just one simple example, add one more command to the above code to cause the cube to start off bright green with each cycle, and to shift smoothly to bright red:

```
color([$t,(1-$t),0])
translate([$t * 20,0,0])
cube(5);
```

To the right of the three number fields is a check box labeled "Dump Pictures". Check this during a cycle or two to cause each frame of the animation to be saved to a file in the same location as the project file. There are ways to convert these images into an animated GIF file, or to use in other ways, but that's beyond the scope of this book.

Notes

Index

$fn - recipe 1, 8
$t - appendix A
Bold - recipe 12
Boolean operators - recipe 14
Box - recipe 16
Cake - recipe 12, 19, 20
Candle - recipe 15
Cartesian - recipe 12
Center - recipe 2, 9, 16
Chamfer - recipe 7
Characters - recipe 8
Cheatsheet - recipe 8
Circle - recipe 4, 10
Color - recipe 10, 12, 17
Comment lines - recipe 6
Cone - recipe 15
Cube - recipe 11, 15, 16, 18
Cylinder - recipe 14, 15, 21
Delta - recipe 7
Difference - recipe 10, 14, 15, 16, 22
Doughnut - recipe 10, 13
Echo - recipe 11
Ellipsoid - recipe 18
Equilateral triangle - recipe 12
Extrude - recipe 9
Faces - recipe 17
Font - recipe 12
For - recipe 11
Function - recipe 12

Happy birthday - recipe 12
Height - recipe 15
Hexagon - recipe 12
Holey grail - recipe 14
Hull - recipe 21
Inches - recipe 16
Include - recipe 19
Intersection - recipe 14
Intersection_for - recipe 11
Italics - recipe 12
Lid - recipe 16
Linear_extrude - recipe 9, 12
List - recipe 2, 11
Math - recipe 11
Millimeters - recipe 16
Minkowski - recipe 22
Mirror - recipe 20
Module - recipe 12
Normal - recipe 20
Octagon - recipe 12
Offset - recipe 7
Order of commands - recipe 5
Origin - recipe 20
Parameterization - recipe 3
Parameters - recipe 3
Pentagon - recipe 12
Perpendicular - recipe 20
Pi - recipe 17
Pie - recipe 17
Pipe - recipe 15
Plane - recipe 20
Points - recipe 17
Polar - recipe 12
Polygon - recipe 6, 10, 14
Polyhedron - recipe 15, 17
Projection - recipe 9, 13

Radius - recipe 12, 15
Range - recipe 11
Rectangle - recipe 2
Rectangular solid - recipe 16
Reflection - recipe 20
Regular_polygon - recipe 12
Resize - recipe 18
Reverse - recipe 20
Right-hand rule - recipe 5, 17
Rotate - recipe 5
Rotate-extrude - recipe 10, 14
Round_edges_box - recipe 21
Rounded_box - recipe 21
Salmon - recipe 13
Scale - recipe 9, 12, 18
Scoop - recipe 20
Shrink-wrap - recipe 21
Slices - recipe 9
Space points - recipe 17
Sphere - recipe 1, 21
Square - recipe 2, 10
STL - recipe 16
STL units - recipe 16
Straw - recipe 15
Strings - recipe 8
Taper - recipe 15
Text - recipe 8, 12
Translate - recipe 3, 5, 19
Triangle - recipe 6, 12
Tube - recipe 15
Twist - recipe 9
Unicode - recipe 12
Union - recipe 12, 14, 16
Use - recipe 19
Variables - recipe 11
Vector - recipe 2

Notes

About John Clark Craig
(johnclarkcraig.com)

John Clark Craig is an author, inventor, and Maker at heart. John has authored about two dozen popular books, published by Microsoft Press, O'Reilly Media, and others, covering several programming languages and technical topics ranging from Visual Basic to high accuracy Sun Position algorithms presented in nine different programming languages.

John used 3D modeling to create parts for his two consumer electronics inventions, a wires-free automatic brake light for bicycles called Lucidbrake, and a hand-held signalling device called LucidLights that lets users create words and graphics in thin air using persistence of vision. John used expensive commercial design programs for the plastic parts, but wishes like crazy he had discovered OpenSCAD a little earlier!

John is also passionate about solar, wind, and other alternative energy sources. He designed and created all software for several of the world's largest (at the time) solar energy heliostat and photovoltaic tracking fields, and he created software to help engineers design some of the world's tallest wind turbine towers.

John lives in, and greatly enjoys, Colorado with his wife and fellow entrepreneur Ellen, and his extended family.

To contact the author
and see more examples of
OpenSCAD, visit:

OpenSCADbook.com

or write to John directly at:
John@OpenSCADbook.com

Made in the USA
San Bernardino, CA
13 April 2019